TIPS ON HOW TO SUCCEED IN MATHEMATICS

(an eye opener to success)

Prof. Augustus N. Wali

PART ONE

PART TWO

PART THREE

PART FOUR

PREFACE

This book is presented in an easy, simple and understandable manner. The flow of information is arranged in a systematic and logical way with some stories and wise quotes appearing to back up the information. Am sure you have heard people say that mathematics is a hard and difficult subject. But, mathematics is not as difficult as people think.

"If people do not believe that mathematics is simple, it is only because they do not realize how complicated life is" -- John Louis von Neumann.

What matters is the kind of attitude you have towards mathematics. This book aims at demystifying that belief since what you hear is one thing and the action that you take thereafter is another.

To succeed in mathematics, you have to identify your strengths and weaknesses and then try to improve on your weak areas. It has been said that rather than complain about thorns on roses, you should be thankful for roses among thorns.

This book suggests mathematics study strategies that can be applied effectively by students at all levels in order to succeed in mathematics. It makes suggestions on how to develop positive approaches and good mathematics study habits. It is also an excellent, easy and a sure way to a breakthrough and success in mathematics.

The book has been organized into seven chapters:

- The first chapter is introductory and explores the meaning of mathematics and career opportunities for mathematicians.

- The second chapter talks about getting started towards success in mathematics.

- The third explores mathematics study skills and techniques.

- The fourth chapter describes mathematics problem solving strategies.

- The fifth chapter talks about studying for a mathematics test, techniques of final revision and getting assistance.

- The final seventh chapter identifies and discusses some factors that contribute to success in mathematics.

PART ONE

CHAPTER 1

INTRODUCTION

Mathematics is one of the oldest sciences. For many, mathematics has its own intrinsic interest and it is worthwhile studying for its interesting ideas and results. Mathematics has its tentacles in all aspects of life. It can be seen as a circle with its center everywhere without a bounded circumference.

"Go down deep enough into anything and you will find mathematics"

-- Dean Schlicter.

Mathematics is the most extensive field of knowledge and represents a logical approach that can be applied to many different situations in life. Everyday life will be quite difficult if you have no knowledge of mathematics whatsoever.

Mathematics is a language in its own right. It makes use of defined terms, signs and symbolic representations to add precision to communication. It is defined as the science of quantity and space. It is a systematized, organized and exact branch of science. It is the numerical and calculation part of your life and knowledge. It helps you to give exact interpretations to your various ideas and conclusions. It is the science which uses easy words for hard ideas.

Mathematics does not present isolated skills and procedures. It is a science of logical reasoning and numerical problems. Mathematics is everywhere and most of what you see is a combination of different concepts.

Mathematics is the art of saying the same thing in many different ways. It is a way to settle in the mind a habit of reasoning. It is exact, true and to-the-point knowledge and therefore creates a discipline in the mind.

Mathematics is a bit tricky: even when it looks as if it is what you expect it to be, you turn your back for a moment and it is changed. Mathematics is a science of all sciences and art of all arts. It is the pivot of all the sciences and arts. It is the door and key to the sciences.

"Mathematics is the supreme judge; from its decisions there is no appeal"

-- Tobias Dantzig.

Mathematics uses penetrating techniques of thought that you can use to solve problems, analyze situations and sharpen the way you look at your world. Most importantly, mathematics relates to things that you do in the real world every day.

In today's world, you are bombarded with data that must be absorbed, sorted, organized and used to make decisions. The underpinnings of everyday life, such as making purchases, choosing insurance or health plans and planning for retirement, all require mathematical competence.

Business and industry need workers who can solve real-world problems, explain their thinking to others, identify and analyze trends in data and use modern technology. Therefore, mathematics relates to other subjects and its study can lead to a variety of exciting professional careers. Basic research, engineering, finance, business, and government services are among the opportunities open to those with mathematical training.

Moreover, with the increasing importance of basic science and information technology, prospects for careers in the mathematical sciences are very good.

Mathematical analysis and computational modeling are important for solving some of the most pressing problems, for instance, new energy resources, climate change, risk management, epidemiology, just to name a few. You must strive to maintain your technological edge and so mathematical skills are crucial to this effort. Some more specific business positions include portfolio analysis, design studies, statistical analysis, computer simulation, software design and testing and other areas of operations research.

There are extensive opportunities for mathematics in finance, the actuarial fields and economic forecasting. Many laboratories, both government and private, maintain independent research staffs that include mathematicians. Their work often deals with the development of new technology, including research in basic physics and software development as well as applied mathematics.

Practical considerations aside, there is the pleasure of learning, applying and creating mathematics. Real world issues pose problems that can be studied by formulating and analyzing mathematical models. In some cases applications may lead to new mathematics and a new branch of science is born. In other cases abstract theory finds unexpected practical purpose. Working on research problems is exciting; solving difficult problems successfully, for many, provides enough satisfaction.

While a career in mathematics can be very attractive, it takes time to acquire the necessary skills. The course sequence provides a foundation upon which more advanced mathematics will be built, while the breadth and depth of work depends on the mathematics study level.

Studying mathematics develops a number of important skills for solving problems suggested either by mathematics or from real world challenges.

"For the things of this world cannot be made known without knowledge of mathematics"

-- Roger Bacon.

You can also study mathematics for the purpose of developing your mind, that is, in developing your ideas and abilities.

Mathematics is part of the culture of your time and you need some understanding of it. You cannot do without the use of fundamental processes of mathematics in your daily life. A common man can get on sometimes very well without learning how to read and write but can never get on easily without learning how to count and calculate. Any person ignorant of mathematics will be at the mercy of others and will be easily cheated.

The knowledge of mathematical fundamental processes and the skill to use them is the preliminary requirement of any human being these days. Even a laborer has got to calculate his/her wages, make individual purchases from the market and adjust the expenditure to his/her income. Although there are many applications of mathematics, the questions that the subject of mathematics seeks to resolve are questions whose answers deepen our understanding of, and reveal new insights into, mathematics itself.

CHAPTER 2

GETTING STARTED/ORGANIZED

(The secret of getting ahead is getting started)

2.1 Creating a Good Study Environment

Students who find school work easier are those who can organize their work. They are able to work on their own, fit their work into a flexible timetable and are willing to stick to it. You have to create a good study environment that allows you to maximize the learning efficiency. When combined with effective time management, high motivation, good reading & note taking skills and systematic test preparation, a good study environment serves as a catalyst for productive effort.

A good study environment is a highly individualized matter. What is right for one student may not be right for a friend or roommate. As much as possible, you should designate a special place to study. This place should be neat and should provide few distractions to allow for maximum concentration. Needless to say, some study tasks must be done elsewhere, but having a regular "home base" that you associate with studying helps to reinforce your self-discipline.

There are three primary considerations in creating your study environment, namely, learning style, learning task and availability of learning resources. In determining your learning style, you must pay attention to your senses. You should eliminate things in your environment that interfere with your concentration and utilize your senses to accommodate learning. You must consider the following: What level and what kind of noise can you tolerate? Do you work best in total silence or with a low to moderate noise level? You may not be able to concentrate well with a lot of noise right outside your window.

Adequate lighting is a must and the direction and intensity of the lighting are important factors. Poor lighting is a chief cause of eyestrain and headaches among students and improper lighting also leads to fatigue. Coverage of study materials should be even with no shadows or glare. Shielded full-spectrum fluorescents may cause you to be calmer, steadier and less easily distracted.

Another "sight" consideration is the level of movement in your study area. Studying near a hallway in the dormitory or an aisle in the library can lead to distractions in watching the passing scenery. Your environment should be comfortable but not too comfortable. Slight amounts of muscular tension have been found to increase efficiency and accuracy in mental work. A chair which promotes good posture is a better choice than a recliner.

You must be aware of appealing aromas (such as the smell of food coming from a nearby kitchen or exotic perfumes worn by the opposite sex) in the immediate vicinity of your study environment. Such distractions, though pleasant, may interfere with concentration. To avoid them, you should move to another area and allow for more efficient use of time.

The nature of the learning task sometimes dictates a particular study environment. If heavy-duty memory work is called for, you may want to study alone for a while and then get together with someone else for a recall drill. For problem-solving, a study group may be your choice provided the group sets some ground rules about staying on course. This usually works best when the group members have done preliminary studying before the group convenes.

You should accumulate the necessary resources like textbooks, notebooks, pens, etc. before you begin to study. You must plan ahead if an assignment requires the use of a textbook on reserve at the library. You should utilize available facilities when appropriate. Speaking of resources, do you find yourself with a stock of junk food on hand when studying? Constant nibbling is not only bad for your waistline but also distracts you from studying. Try having fresh fruits and vegetables, but save munching time for a study break.

There is no one best study environment. You determine your best place by being aware of your learning style, the learning tasks, and the required learning resources. You should have one regular place to study which offers minimum distractions.

2.2 Time Management

Wise use of time leads to success.

"We all sorely complain of the shortness of time, and yet have much more than we know what to do with. Our lives are either spent in doing nothing at all, or in doing nothing to the purpose, or in doing nothing that we ought to do. We are always complaining that our days are few, and acting as though there would be no end of them"

-- Seneca (BC 3-65 AD).

There are, of course, 168 hours in a week. Therefore, you can divide that time between

(i) important and unimportant tasks

(ii) useful and useless tasks

(iii) productive and unproductive tasks.

But, the first task you have is to find out where, when, how and why time is being wasted. Remember, how and what you do is up to you, no-one will ever do it for you. It is important to note that, if you fail to plan then you are planning to fail. You must think carefully and be honest with yourself.

"Don't be fooled by the calendar. There are only as many days in the year as you make use of. One man gets only a week's value out of a year, while another gets a full year's value out of a week"

-- Charles Richards.

Mostly, it is recommended that you can plan to spend at least two hours outside of class for every hour spent in class. Therefore, if we are taking 15 hours a week in class, it is suggested that we spend at least 30 hours a week outside of class studying and doing assignments.

2.3 Long Term Planning

Planning gives a clear sense of direction and purpose. Direction and purpose make your study more meaningful. This builds motivation and self-confidence, which you can direct to achieve your goals. Long term planning could mean: what you intend to do in a week's time, to what you intend to do with the rest of your life.

Many students fail tests or examinations because they start their revision too late rather than because they lack ability. The reason for this seems to arise from the many other pressures put upon them. Friends, family and other responsibilities all eat away at the valuable time. No wonder revision gets put off to just before a test or examination. Strange as it may seem, careful planning of what you do and when to do it often uncovers huge amounts of free time you never knew you had. You should remember to get organized early and revise regularly.

2.4 Constructing Timetable

There are several working weeks in a typical school year. You can use this as a guide for your rough timetable and fill in tasks as they are given including the times tests take place, other given deadlines and vacations. This has an advantage for it shows at a glance timings for every activity. Busy periods, lead-in time to deadlines, show up clearly and time limits for tasks can be set and adhered to. The timetable should then be backed-up with a detailed daily plan which states what you need to do day by day.

2.5 Daily Timetabling

It is said that, today's preparation determines tomorrow's achievement. Therefore, you should plan your activities for each day. While planning, you must not forget to include breaks, meal times and leisure activities. You should also try to be as accurate about the time needed as you can. For the evening, you can make up a timetable from the time you get home and divide your time in slots of 15 minutes or so.

Most of the weekends will be taken up with other things but you must try to spare some time to prepare for the week ahead and/or revise for tests and examinations.

You should remember that everyone is different. The timetable must suit you and not your needs or what someone else says you must do. You should never forget what your timetable is for.

2.6 Working Out Priorities

Priorities can be divided up into:

(a) Urgent--must be done now

(b) Important--must be done soon

(c) Unimportant--must be done eventually.

There is a lot of overlap in doing this, so you might list the things you have to do only in order of importance--the top one or the top two becoming urgent. You need to keep a diary and a pencil handy to record things as they crop up and be flexible. Another way of setting priorities is to use the 4D system:

(i) Dump anything that does not need to be done at all

(ii) Delay what you cannot dump

(iii) Designate a time for what you cannot delay and then

(iv) Do it.

If you follow this carefully, the things you might see as difficulties will become enjoyable challenges.

2.7 Getting The Most Out Of Learning Abilities

The only way to have a good mathematics study time is to avoid boredom. Boredom is one of the worst enemies of learning. From a very early age, you varied what you did to stop yourself from getting bored. You looked for new tasks and sought new ways to do things. Sometimes even this failed and you fell into the boredom-trap. At this stage, you cannot be bothered to find ways to cure your boredom because you are too bored. Over time you developed a concentration-span, that is, the time between starting a task to the time you find your mind wandering. This is because your brain deals with information in a very special way. Your brain receives information from your senses. This is then passed on to your short-term memory, which has limited capacities, where it is stored for a short time. From then on the information in short-term memory must be passed on to long-term memory, that has an unlimited capacity, or else you end up forgetting the information.

Your brain takes in much more information than you need. If you do not make active use of the information, the path to it becomes lost or overgrown, making it hard to access. The way you take in information also affects what you can remember. Your brain's ability to do this depends on how you feel. If you are bored, short-term memory losses information very quickly and so can never be passed on. The information is not lost forever, it is stored away never to be remembered again.

The best way to learn mathematics is to limit study periods to the length of your concentration-span. This gives the brain the best chance to store what you are studying in the long-term memory. You should never study beyond your concentration-span, since as you continue reading your brain will be losing most of the information it takes in. This makes it pointless to go on. You may satisfy your need to feel as if you are working hard, but the amount you actually remember will get less and less.

After studying for the time that is best for you, you must then take a rest for about five minutes. Do something else not concerned with your work while still thinking about what you have read. Your brain needs that rest time to sort out the information in your short-term memory. At the end of the rest period, the information you have read will be clearer than it was to begin with.

To get the information into long-term memory, you must review. That is, after your five minutes rest, you read the same information again concentrating only on those points that are most important. Then you can take another five minutes break and re-read once more, fitting all the bits of information together.

Both of these reviews would be made even better by note-taking of the basic concepts by whatever well organized way you find easy and helpful. One week and two weeks later you review the topic again using your notes. By then, you will find that there has been a huge improvement in your ability to remember, understand and use that information.

Finally, you must revise. This is simply a way of drawing loose ends together with the same study method but this time using your notes only.

The purpose of notes is not to copy out great chunks of information from books but select out the most important points and concepts. They should only act as a trigger to help you remember what you have read. Key words are more easily remembered than long sentences. Notes should be short, to the point, well-organized and easily read. While revising, you should concentrate only on those pointers that will help you recall the lesson content. Therefore, by improving your reading and note-taking skills, you can increase the ability to remember.

What you already know and have a name for affects what you notice, how you direct your attention and therefore what goes into the memory. You need to maintain your attention in order to remember.

You will remember more if you:

(a) Direct your attention consciously and purposely

(b) Focus in a relaxed way, that is, not with hard concentration. A state of relaxed alertness helps

the imagination and increases suggestibility and openness to new information

(c) Take breaks and make changes in what you are doing, so as to maintain relaxed attention

(d) Link information to what you know

(e) Give names and labels to information.

Flanagan (1997) argues that, you remember: 20% of what you read, 30% of what you hear, 40% of what you see, 50% of what you say, 60% of what you do and 90% of what you read, hear, see, say and do. These are clearly not scientific figures but they suggest the importance of interaction with the material and of using all your senses.

Rehearsing new information in short-term memory helps the working memory hold onto it. Repeating it gives the brain time to call up stored memories to help you make sense of the information and encode it for storage. Rehearsal must start within a few seconds, as information fades quickly. Rehearsal is a useful strategy for holding onto numbers, formulae and instructions long enough to write them down.

The brain encodes new information so that it can be represented in the memory. For example, when you tell a story, the brain encodes the pattern of fine-muscle movements you used to speak and stores them. It can also encode and store the sound of your speech on your own ear; the images and emotions which the story brought to mind; the look of the text; and details such as who was in the room. The brain links information it has encoded so that any one aspect could trigger the whole memory later. The more facets of an experience the brain has encoded, the more triggers to memory.

It follows that; you can assist your memory by choosing to encode information in several ways such as;

(i) Use of a different room for each topic or associate furniture, windows and plants with

particular topics.

(ii) Use of parts of your body as triggers to memory, as your body will be there in the

examination room. For example, each finger could represent one major topic; each segment

of each finger a principal mathematical concept you would use etc.

(iii) Study on the move. When you exercise, you can associate each movement with something

you wish to remember. To refresh the memory, you go through the exercise in your mind.

(iv) Go over a topic with a real or imaginary friend.

(v) Make page layouts clear and attractive.

(vi) Assign to a topic an object and label different bits of the object with the things you need to

remember.

(vii) Spend time considering the implications of what you have found out. Think of a different

way of saying what you have already written.

Good recall is linked to how much attention and awareness you bring to the process of taking in the information and encoding it. The more visual you can make the learning process, the easier it will be to recall the information. You must be creative and use your imagination. You should create mental images that you can associate with the information you are trying to learn. If you want to recall information at will, such as for examinations or for complicated sequences that you use regularly, you can have fun with the memory process and play with information until you find a helpful mnemonic. One common mnemonic is to use the first letter of each keyword to make a new word. For example, while studying trigonometric ratios topic, you can put the letters A, S, T and C in the four quadrants of a unit circle to stand for: all angles are positive in first quadrant, only sine is positive in second quadrant, only tangent is positive in the third quadrant and only cosine is positive in the fourth quadrant respectively. To remember this you can use the mnemonic, All Students Take Coffee or All Smokers Take Care, to remember the signs in the four quadrants. By combining all of these strategies, you can greatly enhance your memory potential.

PART TWO

CHAPTER 3

MATHEMATICS STUDY SKILLS

First and foremost, everyone studies differently and there is no right way to study mathematics. You study differently and ought to do the best that you can. A skill, in general, is a learned activity developed through practice and reflection. To develop a skill you need first to know where to start from, what your current strengths and weaknesses are, what you want to achieve or improve and what could obstruct your goals. Mathematics study skills are part of a developmental process. They evolve and mature through practice, trial and error, feedback from others and reflection as you move through the different stages of the course. However, there are some basic approaches which can start you off on a good footing, help you cut corners and accelerate the learning process. Most important in the development of mathematics study skills, is to ensure that your work is carried out steadily and regularly.

The tools for reading a mathematics book are pen/pencil and paper. In most mathematics textbooks one line may follow from another, but often intermediate steps are left out, and that is why you need a pen and paper, to try and provide the missing steps. The author often assumes that you are able to fill in the missing steps, particularly if something similar had been explained before.

"Arithmetic is numbers you squeeze from your head to your hand to your pencil/pen to your paper till you get the answer"

-- Carl Sandburg.

To enhance your mathematics study skills, you have to be involved in managing the learning process, the mathematics and your study time:

(a) You have to take responsibility for studying, recognizing what you do and do not know, and knowing how to get your teacher to help you with what you do not understand.

(b) It is imperative that you attend class every day and take complete notes. Teachers formulate test questions based on materials and examples covered in class as well as on those in the textbook.

(c) You are supposed to be an active participant in the classroom. It is good to get ahead in the textbook by trying to work some of the problems before they are covered in the class.

(d) You must be able to ask questions in the class. There are usually other students wanting to know the answers to the same questions you have. Remember, a person who asks a question is a fool for 5 minutes; a person who never asks a question is a fool for the rest of his/her life.

(e) When faced with problems that you are unable to solve, it is advisable that you seek help from your teacher or other students. The teacher will be pleased to see that you are interested and are actively helping yourself.

You must remember that studying mathematics is different from studying other subjects. Therefore, you should be prepared to learn mathematics by doing many problems. So you should do your homework. The problems that you solve help you to learn the formulas, concepts and techniques that you need to know as well as improve your problem–solving prowess (skill).

"The essence of mathematics is not to make simple things complicated, but to make complicated things simple"

You cannot learn mathematics by just going to class and watching the teacher teach and work problems. In order to learn mathematics you must be actively involved in the learning process. Since each class builds on the previous ones, you must keep up with the teacher by attending class, reading the textbook and doing your homework every day. Falling a day behind puts you at a disadvantage while falling a week behind will put you in deep trouble.

While studying mathematics you are always reviewing previous material as you do new material. This makes many of the ideas hang together and helps in identifying and learning the key concepts. You should remember that mathematics is cumulative. If you do not go to class you will miss important material that will be used in later sections.

In higher level school mathematics you are expected to learn and absorb new material much more quickly. Tests are probably spaced farther apart and so cover more material than before. The teacher may not even check your homework.

Therefore;

(i)   You should take responsibility for keeping up with the homework and make sure you find out how to do it. After the end of each class you must budget some time to look over the homework from that day's lesson and attempt to do it while the lesson is still fresh in your mind. Doing this will allow you time to really work at understanding the concepts covered that day. You should not wait until the last minute to do the homework as this often results in an incomplete homework set and/or an incomplete understanding of the concepts.

(ii)   You probably need more time studying per week by doing more of the learning outside of class than in lower level school mathematics.

(iii) Tests at higher level school mathematics may seem harder just because they cover more material.

3.1 Study Time

You may know a rule of thumb about mathematics classes: at least 2 hours of study time per class hour. But this may not be enough.

Thus;

(a) You need to take as much time as you need to do all the homework and to get complete understanding of the material. Homework utilizes your leisure time which otherwise would have been wasted and thus it may establish the habit of working hard. In mathematics, self-study is essential to ensure steady progress. The work cannot be completed without doing a part of it as homework. The efficiency in solving problems has to be developed by you solving them. You must practice as much as possible. The only way to learn how to do mathematics problems is to work out lots of them. The more you work, the better prepared you will be, come examination time.

(b) The more challenging the material, the more time should be spend on it.

(c) You should also form a study group. A study group:

(i)   Reinforces, clarifies, and deepens your learning by providing the opportunity to teach one another. Many students improve their grades by supplementing individual study with group study.

(ii) Provides feedback before the test on how well you are learning the material.

(iii) Prepares you for the working world, with its emphasis on teamwork.

(iv) Provides a "support group." All students feel discouraged at times, but a study group can "refuel" your motivation and make studying more fun.

(vi) Helps you overcome shyness about discussing issues in class.

(vii) Helps you become motivated to study, because you know your study group is depending on your preparation.

A man died and St. Peter asked him if he would like to go to heaven or hell. The man asked if he could see both before deciding. St. Peter took him to hell first and the man saw a big hall with a long table, lots of food on the table and sweet music playing in the backyard. He also saw rows of people with pale, sad faces. They looked starved and there was no laughter. And he observed one more thing. Their hands were tied to four-feet spoons and they were trying to get the food from the center of the table to put into their mouths. But they could not. Then, he went to see heaven. There he saw the same scenario. But he observed there was something different there. People were laughing and were well fed and healthy-looking. He noticed that they were feeding one another across the table. The result was happiness, enjoyment and gratification because they were not thinking of themselves alone; they were thinking win/win.

The moral of this story is, together each achieves more (T.E.A.M). Teamwork divides the task and doubles the success.

"Coming together is a beginning. Keeping together is progress. Working together is success"

-- Henry Ford.

After forming a study group you should meet once or twice a week and go over mathematics problems that you have trouble with. During the meeting, either someone else in the group will help you, or you will discover you are all stuck on the same problems. Then it is time to get help from your teacher.

3.2 Methods/Techniques Of Studying Mathematics

The following are some of the methods/techniques of studying mathematics that you can employ in order to succeed in mathematics. It is important not to think of any of them as a magic formula that is the passport to success. You should not apply them blindly, for no technique is useful 100%. Different study methods work best in different situations.

3.2.1 Cornell Method

When using this method, you divide the pages of your notebook into a narrow column on the left and a wide column on the right. The right hand column is for detailed notes. The left, which is headed recall, contains one or two key words which summarize the corresponding material in the right hand column.

When studying the notes, you cover up the right hand column and use the left hand column to prompt you into recalling the material in the right hand column. When discussing the material with fellow students and during revision, you can use this method.

3.2.2 SQERPSRR method

In this method:

(a)   **S** stands for survey or skim (read quickly). Before beginning to read you must look through the

whole chapter, read the chapter title, introduction, learning objectives and end of chapter

questions. You must also note the main definitions and concepts. This step helps you to gather

the information necessary to focus on the chapter, create a mental framework for understanding

and formulating questions for yourself as you read the chapter.

It is not necessary to have answers to your questions at this step of the process; the answers will come later. You should also note the organization of the chapter or textbook you are studying, the main concepts and try to get some idea of methods of proof.

(b)   **Q** stands for questions. Now that you have surveyed the entire chapter to build a framework for understanding, it is time to begin the reading process. You can try to turn each heading and each subsection into one or more questions and write down the questions on the left third of a piece of paper. For instance, you can ask yourself "what is the point of this or what question do you have that this chapter might help answer?" As you continue reading the section you will be looking for the answers to your questions. When the mind is actively searching for answers to questions, it becomes engaged in the learning process. This helps you to remember and understand the information and creates an active learning environment.

(c)   **E** stands for doing a number of worked examples. You must do all the worked examples in each section making sure that you derive the missing lines. When you do the worked examples, it helps you to acquire skills and master the ideas and concepts.

(d)   **R** stands for careful reading. You read the page or chapter carefully and thoughtfully, making more notes as you go on. You should concentrate on the key words or sections you highlighted, using them as trigger points for recall. Reading the section fills in the information around the mental structures you have been building by surveying the chapter and developing questions about the section. As you read the section you should look for the answers to your questions and jot them down, in your own words, on the right two thirds of your piece of paper.

(e)   **P** stands for problems. At this stage, you should try to solve the problems at the end of the section.

(f)   **S** stands for summary. Now it is the time to go over the text again, writing a summary of the definitions, formulae and main concepts of the section.

(g)   The first **R** stands for recite or rehearse. You can write down a key phrase that sums up the major point of the section. It is important to use your own words, not just copying a phrase from the book. Reciting material as we go on retrains our mind to concentrate and learn as it reads. You must recite at the end of each section of the chapter. You recite by looking at the question(s) you wrote down before you read the section. You then cover the answers with a piece of paper and see if you can answer the questions from memory. If you cannot recall the answers to your questions, you should reread the section or the part of the section that has to do with that question.

(h)   The second **R** stands for review or revision. This is the final step in the reading session. The review step helps you refine your mental organization and begin to build your memory and understanding of the material in the chapter. You learn through repetition. This step provides another opportunity for repetition of the material and therefore will enhance your recall of the information. Once you have finished reading the entire chapter using the survey, question, examples, read, problem, summary and recite steps, you review by going back over all your questions. You cover the answers to the questions you have developed and written down and see if you can still recite them. In case you have forgotten some of the answers, you can reread that section of the chapter to refresh your memory, recite the answers and then continue your review process.

Using SQERPSRR provides a method of reading mathematics that will most likely enhance understanding and retention of material and concepts. It is not a quicker way to read a chapter in a textbook but it is likely to reduce the amount of time needed to spend studying the material immediately prior to a test.

The following are some of the advantages of **SQERPSRR**

(i)   By using this method you realize that it is totally insufficient just to read the textbook and that a large number of steps are required to master the work.

(ii)   This method gives a procedure for coping up with the work. By going through this procedure, you prepare the groundwork for understanding, even if the material does not really make much sense at first sight.

3.2.3 Jig-Saw Puzzle Method

This corresponds to the method of completing jig-saw puzzles; the sort where you are not given the final picture but just the pieces. You assemble little parts of the jig-saw that make sense by themselves and then assemble the completed parts together to form a whole. You get little groups of insights and then build them up to make a whole. It is important to learn not only the facts but to understand how these facts fit into the bigger picture.

3.2.4 City Explorer Method

If you go to a strange city you normally start off with a fixed point; the railway station, bus station or your hotel. Everything else is then related to the fixed point and other points are then related to these new fixed points and so on. While studying mathematics you try to start off with something you know and understand well and then relate new ideas and results to the well-known idea. This is beginning with something you know and moving one step at a time, relating each step to its predecessor.

3.2.5 MURDER method

The acronym (abbreviation) MURDER provides us with a foolproof, systematic approach to studying mathematics.

The respective letters in MURDER stand for:

(a) **Mood:** You should set a positive mood for studying by selecting the appropriate time and environment. Create a work area that is free from distractions and commit to staying there for at least one to two hours. Setting up a study area will help you get into the study habit. If you always work in this one place your brain will begin to switch into study mode as soon as you sit down. In other words you activate your ability to concentrate.

When setting up your workplace you will need to organize your materials so that they are always at hand. There is nothing worse than interrupting study to find something that should have been there already. When searching for materials, you will waste valuable time and your concentration will fade. If you get side-tracked, you have to remind yourself how this activity will help you to meet your goals.

(b) **Understand:** You must mark any information that you do not understand in a particular topic, and then look for it in other textbooks or learning resource materials. You may also ask your teachers or other students to explain it to you. You should also keep your focus on understanding one topic, concept or a manageable group of exercises.

(c) **Recall:** After studying the topic, you should stop and put what you have learned into your own words.

(d) **Digest:** After going through a section you may go back to what you did not understand and reconsider the information. You can also seek help from other books or a teacher.

(e)   **Expand:** To understand the information and concepts fully you may ask three kinds of questions concerning the studied material:

(i)   If you could speak to the author, what questions would you ask or what criticism would you offer?

(ii)   How could you apply this material to other topics or other information you have learned?

(iii) How could you make this information interesting and understandable to other students?

(f)   **Review:** Review is the mental process of going over the material already learnt. It is to view again or to repeat the work. The previous experiences are recalled for better memory. It is meant for better understanding and command. So you must go over the material you have covered. You must also remember the strategies that helped you understand and/or retain information and apply them to your future studying.

CHAPTER 4

PROBLEM SOLVING STRATEGIES

By solving problems, you acquire ways of thinking, habits of persistence and curiosity and confidence in unfamiliar situations that serve you well outside the mathematics classroom. You should have frequent opportunities to formulate, grapple with, and solve complex problems that involve a significant amount of effort. You must explain and discuss your thinking during the problem-solving process so that you can apply and adapt the strategies you develop to other problems and contexts.

When you are challenged to communicate the results of your thinking to others orally or in writing, you learn to be clear, convincing and precise in your use of mathematical language. Your explanations should include mathematical rationales, not just procedural descriptions. Conversations in which mathematical ideas are explored from multiple perspectives help you sharpen your thinking and make connections.

Mathematics is not a collection of separate strands or standards, even though it is often partitioned and presented in this manner. Rather, mathematics is an integrated field of study. When you connect mathematical ideas, your understanding becomes deeper and more lasting and you come to view mathematics as a meaningful whole. You see mathematical connections among different mathematical topics, in contexts that relate mathematics to other subjects and in their own interests and experience.

4.1 Some Problem-Solving Strategies

You need to be able to reflect regularly on your mathematical activities and determine whether you are

making appropriate progress towards solving problems. Just as a carpenter or mechanic makes decisions about which tools to use, you need to reflect regularly on whether the right tool to solve a given problem is being used. If a problem-solving strategy is not successful, you may need to try another strategy, although it is possible that later you might discover that the original strategy was after all appropriate.

There are many reasonable ways to solve mathematics problems. The skill at choosing an appropriate strategy is best learned by solving many problems.

The following are some of the strategies for solving mathematics problems:

(a)   Using one or more variables

(b)   Completing a table

(c)   Considering special cases

(d)   Looking for a pattern

(e)   Guessing and testing

(f)   Drawing a picture or diagram

(g)   Making an orderly list

(h)   Solving a simpler related problem

(i)   Using direct reasoning

(j)   Working backward

(k)   Solving an equation

(l)   Looking for a formula

(m) Using coordinates

(n)   Eliminating possibilities

(o)   Using symmetry

(p)   Using a model

(q)   Using a formula

(r)   Being ingenious

When you create representations to capture mathematical concepts or relationships, you acquire a set of strategies that significantly expand your capacity to model, interpret and analyze physical and social phenomena in a mathematical way.

4.2 Problem Solving

Problem solving is the key in being able to do all other aspects of mathematics. Through problem solving, you learn that there are many different ways to solve a problem. It involves the ability to explore, think through an issue and reason logically to solve routine as well as non-routine problems. In addition to helping with mathematical thinking, this activity builds language and social skills such as working together. It creates means of finding ways to express ideas with words, diagrams, pictures and symbols.

Before attempting to solve a problem, you must understand it. You have to do a bit of reasoning. Success in solving mathematical problems is not possible without the use of creative powers of the mind.

Reasoning is used to think through the problem and come up with a useful answer. This is a major part of problem solving. You should do some calculations, which enable you to get a sense of the problem. You can also draw a graph or a diagram.

"Haven't I seen you somewhere before?" This is a comment useful at a party to try to meet somebody you fancy. But it can be used effectively in mathematics to find out what is relevant to the problem at hand. You can try to look for a definition or an example in a textbook, which looks similar. While doing this, you can try to learn these relevant definitions and examples. If you do that, at least you accomplish something, even if you have not solved the problem.

For instance, to solve for x and y in the two equations $ax + by = c$ and $dx + ey = f$ where a, b, c, d, e and f are given constants, you ask yourself: where have i seen something like this before? The answer could be "from the topic of solving simultaneous equations." Then you ask: what do you know about simultaneous equations? The answer is: these are equations that are solved at the same time. Do you know any method(s) of solving simultaneous equations and if so what is/are the method(s)? The answer is: the methods of solving a simultaneous equation are (i) substitution method (ii) elimination method (iii) graphical method (iv) matrix method or (v) Cramer's rule method. Is there any specified method in the question to use in solving the problem? If not then you select any of the methods bearing in mind to pick the one in which you can use the shortest time possible in case you are doing an examination. All the methods given for solving a simultaneous equation lead to the same answer/solution.

When you work problems on homework, you must write out complete solutions as if you are taking a test. You should not just scratch out a few lines and check the answer in the back of the book. If your answer is not right, you must rework the problem. You should not just do some mental gymnastics to convince yourself that you can get the correct answer. If you cannot get the answer, you must get help. Having completed a problem or proof, you should examine it again and check whether you can understand it better or whether there is anything else you can learn from it. The practice you get by doing homework and reviewing will make test problems easier to tackle.

In his book entitled "How to solve it", Polya (1945) subdivided problem solving into four steps:

(1) understanding the problem,

(2) devising a plan,

(3) carrying out the plan, and

(4) looking back.

The steps outline a series of general questions that you can use to successfully write resolutions to a problem. Without the questions, common sense goes through the same process; the questions simply allow you to see the process on paper. The questions are designed to be general enough to be applied to almost any problem.

(1) **Understanding the Problem** (Recognizing what is asked for)

In this first step, you should be able to state the unknown, or the thing you want to find to answer the question, the data the question gives you to work with and the condition, or limiting circumstances you must work around. If you can identify all of these and explain the question to other people, then you have a good understanding of what the problem is asking.

You can draw a picture if possible or introduce some kind of notation to visualize the question. This seems so obvious that it is often not even mentioned, yet you are often hindered in your efforts to solve problems simply because you do not understand it fully or even in part.

To understand a problem fully, you must ask questions such as:

    a.   Do you understand all the words used in stating the problem?

    b.   What are you asked to find or show?

    c.   Can you restate the problem in your own words?

    d.   Can you think of a picture or a diagram that might help you understand the problem?

    e.   Is there enough information to enable you to find a solution?

(2) **Devising a plan** (Responding to what is asked for)

By devising a plan, you can eliminate mistakes that you might make by rushing into the actual execution of the plan. This is done by identifying which skills and techniques can be applied to solve the problem at hand. When you plan it out first and then do the mathematics, it is possible to check your work as you go along.

To devise a plan, you can start by trying to think of a related problem you solved before to help you. If you can think of a problem you solved before that had a similar unknown, it could also be helpful. You can also try to restate the problem in an easier or different way and try to solve it. By looking at these related problems, you may be able to use the same method or other part of the plan used. After you have decided which calculations, computations or constructions that you need and have made sure that all data and conditions are used, you can try out your plan.

(3) **Carrying out the plan** (Developing the result of the response)

This step is usually easier than devising the plan. In general, all you need is care and patience given that you have the necessary skills. You must persist with the plan that you have chosen. If it continues not to work you should discard it and choose another. To carry out the plan, you must do all the calculations and check them as you go along. Then you should ask yourself, "Can you see that it is right?" and then, "Can you prove it is right?"

(4) **Looking back** (Checking. What does the result tell you?)

Much can be gained by taking the time to reflect and look back at what you have done, what worked and what did not. Doing this will enable you to predict what strategy to use to solve future problems. When you look back on the problem and the plan you carried out, you can increase your understanding of the solution. It is always good to recheck the result and argument used and to make sure that it is possible to check them. Then you should ask, "Can you get the result in another different way?" and "Can you use this for another problem?"

The solution to question (a) below, shows the incorporation of these steps in forming a logically sound and reliable problem solving process. The next two problems (b) and (c) are solved, indicating the mental questions you ask yourself.

(a) If Tom has three times as many oranges as Susan and Susan has one-fourth as many as John, find how many oranges Mary has given that she has two more than Tom and that John has four.

Solution: After understanding the problem, you respond to what is asked for. This is done by developing the plan. Thus you have:

$$\text{Mary} = \text{Tom} + 2 \dots\dots\dots\dots\dots\dots\dots\dots\dots\dots\dots\dots \text{Plan 1}$$

$$\text{Tom} = 3*\text{Susan} \dots\dots\dots\dots\dots\dots\dots\dots\dots\dots \text{Plan 2}$$

$$\text{Susan} = \text{John}/4 \dots\dots\dots\dots\dots\dots\dots \text{Plan 3}$$

$$\text{John} = 4 \dots\dots\dots\dots\dots\dots \text{Plan 4}$$

Plan 1 is devised for responding to what the problem asks for. In this case, the problem asks for the number of oranges Mary has. The information for formulating the plan is in the problem statement.

Plan 2 is the plan devised to respond to what Plan 1 asks for. The information for Plan 2 is contained in the problem statement.

Plan 3 is the plan devised to respond to what Plan 2 asks for. You repeat the process until there are no more requests. With the devising of the plan complete, you then carry out the plan.

This consists of successive substitutions to obtain:

Susan = 1…………………………..Carrying out plan 3

Tom = 3…………………………...Carrying out plan 2

Mary = 5…………………………..Carrying out plan 1

Hence, Mary has 5 oranges.

It is to be noted that each of the plans is a new problem independent of the previous plan. Understanding the problem is applied to the plan last written to determine what that plan asks for. The planning of the solution is the thinking part. In general, it requires use of knowledge. In simple problems like this one, the knowledge is contained in the problem statement.

(b) If $\frac{a}{b} = \frac{c}{d}$ prove that $\frac{ac-2b^2}{b} = \frac{c^2-2bd}{d}$

Solution: In this case, the equation $\frac{ac-2b^2}{b} = \frac{c^2-2bd}{d}$ is to be proved true given that $\frac{a}{b} = \frac{c}{d}$.

You ask yourself: what should be the first step towards the simplification of these two sides of the equation? (Cross multiplication).

Therefore, using cross multiplication, $\dfrac{ac-2b^2}{b} = \dfrac{c^2-2bd}{d}$ will be true if $acd - 2b^2d = bc^2 - 2b^2d$

What is the next possibility of further simplification? (Addition of $2b^2d$ to both sides of the equation).

Thus $acd - 2b^2d = bc^2 - 2b^2d$ will be true if $acd = bc^2$

What next plan can be carried out? (You divide both sides of the equation by c).

Hence $ad = bc$ will be true if $\dfrac{a}{b} = \dfrac{c}{d}$, which is known and true. Therefore by going back through the chain of arguments, you can say that $\dfrac{ac-2b^2}{b} = \dfrac{c^2-2bd}{d}$ is also true.

(c) A side of a triangle is produced so as to form an exterior angle. Show that the sum of the interior opposite angles is equal to the formed exterior angle.

Solution: You ask yourself: what is given in the problem?

(A triangle is given whose one side is produced so that an exterior angle is formed). Using drawing skills you draw a triangle ABC with BC produced to D, so that an exterior angle ACD is formed.

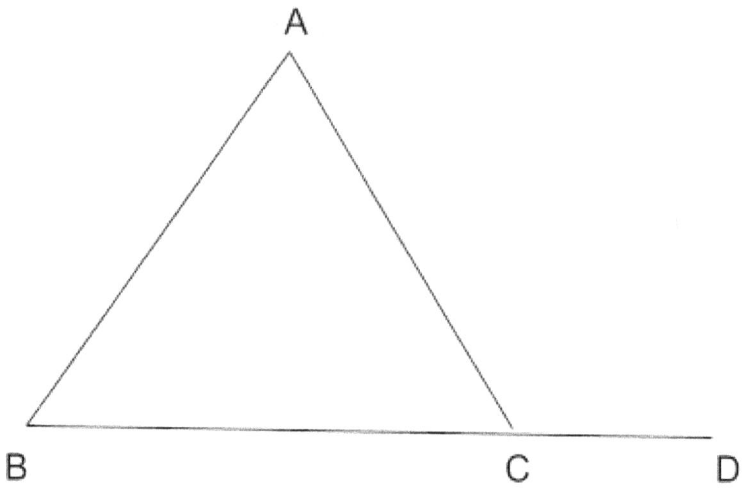

Next, you ask: what has to be shown? (You are to show that $\angle ABC + \angle BAC = \angle ACD$, that is, the sum of the two opposite interior angles equals to the exterior angle).

You also ask yourself: what is the sum of the three interior angles of a triangle? (It is equal to 180°).

Thus you have,

$$\angle ABC + \angle BAC + \angle ACB = 180^\circ \qquad (1)$$

You again ask whether there is another group of angles in the diagram whose sum may be equal to $180°$.

You know that a straight line angle is $180°$. Thus

$$\angle ACB + \angle ACD = 180° \qquad (2)$$

Where do these two equations (1) and (2) lead you to? (You can say that, since each of them equals to $180°$ then they must be equal). Therefore,

$$\angle ABC + \angle BAC + \angle ACB = \angle ACB + \angle ACD$$

What can you conclude further? ($\angle ACB$ is common and thus can be subtracted from both sides). Hence you get,

$$\angle ABC + \angle BAC = \angle ACD.$$

This finalizes the proof.

The mental questions in each step facilitate understanding. They also strengthen the urge to discover facts. The steps are developed in a general manner. No cramming of a fixed step and a set pattern is necessitated. Each step has its reason and justification. Thus you grapple with the problem confidently and intelligently gaining comprehension and problem solving skills.

In early mathematics courses, you solve problems testing memorization, skills and requiring application of skills to familiar situations. Such problems often require just one step to find a solution. Later courses, expect you to tackle more problems requiring application of skills to both familiar and unfamiliar situations and (eventually) requiring that you extend the skills or theories that you know before applying them to unfamiliar situations. Each problem requiring application of skills and extension of the skills or theories that you know before applying them to unfamiliar situations usually requires you to use a multi-step approach and may involve several different mathematical skills and techniques. When faced with such problems, you have to break them down into smaller pieces and solve each piece. (Divide and conquer rule).

The following are some types of such problems that you encounter:

(a)   **Problems testing memorization**

For instance; (i) given that a square has an area of sixteen square centimeters, what is the length of each of its sides?

Solution: In this case, you are expected to know the formula for the area of a square, that is, the formula for the area A of a square with side-length s is:

$$A = s^2.$$

Since you are given the area, you substitute this value into the area formula to get:

$$16 = s^2.$$

Solving this for s you have:

$$4 = s.$$

Thus, the length of each side is 4 centimeters.

(ii) In a group of students, 20% failed a mathematics examination, 10% failed a physics examination and 7% failed both in mathematics and physics. What is the probability that a student (selected at random) who fails in physics also fails in mathematics?

Solution: In this case you are expected to know that, in general, given two events E_1 and E_2 in the same probability space, the conditional probability of E_1 given E_2 that is, the probability that event E_1 occurs given that event E_2 has already occurred, is defined by

$$P(E_1/E_2) = \frac{P(E_1 \cap E_2)}{P(E_2)},$$

where, it is assumed that E_2 is not impossible, that is, $P(E_2) > 0$.

So, let E_1 be the event that a student fails in mathematics and E_2 that of failing in physics. Then,

The probability that a student fails in mathematics is,

$$P(E_1) = \frac{20}{100} = 0.2,$$

the probability that a student fails in physics is,

$$P(E_2) = \frac{10}{100} = 0.1$$

and, the probability that a student fails in both mathematics and physics is,

$$P(E_1 \cap E_2) = \frac{7}{100} = 0.07$$

Hence, the probability that a student who fails in physics also fails in mathematics is

$$P(E_1/E_2) = \frac{0.07}{0.1} = 0.7$$

(b) **Problems testing skills**

For example; (i) assume you have nine identical-looking stones in which gold is embedded in one of them. Eight of the stones weigh the same, but the stone containing the jewel weigh slightly more than the others. You have to find the one with gold in it by using a weighing machine that you are allowed to use only twice. Is it possible to select the slightly heavier stone containing the gold from among the nine identical-looking stones? Explain why or why not.

Solution: Initially, you may think that it is impossible to find the jewel since you are allowed to make only two comparisons. Instead of comparing stones individually, you should compare one collection of stones with another collection of stones.

You group the stones into three groups of three and place one group of three on one side of the balance scale and another group of three on the other side. If both sides weigh the same, then you know that the jewel must be in the third group of three. If, however, one side is heavier than the other, then the jewel is in one of the three that weigh more. In either case, after only one weighing, you are able to identify a group of only three stones among which is the jewel.

Next you take two of these three stones and place one on either side of the scale. If one weighs more than the other, then you know that is the stone containing the gold. If they both weigh the same, then you know that the third stone must contain the jewel. Thus, by weighing the stones only twice, you are able to find the jewel.

This example teaches you to take partial steps whenever possible. Notice that, instead of trying to find the jewel immediately, you reduce the number of possibilities from nine to three. Thus you make the problem easier and then crack the easier problem. "Divide and conquer" is an important and useful technique in both mathematics and in life.

(ii) In a group of 50 cars tested by a garage, 18 had faulty tyres, 16 had faulty brakes and 15 exceeded the allowable emission limits. Also, 5 cars had faulty tyres and brakes, 7 failed on tyres and emissions, 8 failed on brakes and emissions and 3 cars were unsatisfactory in all three respects. How many cars had no faults in these three checks?

Solution: To answer this question, let

A = set of cars with faulty tyres

B = set of cars with faulty brakes

C = set of cars with unsatisfactory emissions.

The given information from the problem is

Number of cars with faulty tyres, $n(A) = 18$

Number of cars with faulty brakes, $n(B) = 16$

Number of cars with unsatisfactory emissions, $n(C) = 15$

Number of cars with faulty tyres and faulty brakes, $n(A \cap B) = 5$

Number of cars with faulty tyres and unsatisfactory emissions, $n(A \cap C) = 7$

Number of cars with faulty brakes and unsatisfactory emissions, $n(B \cap C) = 8$

Number of cars with faulty tyres, faulty brakes and unsatisfactory emissions,

$n(A \cap B \cap C) = 3$

The number of cars having at least one of the three faults is

$$n(A \cup B \cup C) = 18 + 16 + 15 - (5 + 7 + 8) + 3 = 32.$$

Let U be the set of all cars tested. Then, the number of all cars tested, $n(U) = 50$.

So, the number of cars with none of the three defects is

$$50 - 32 = 18.$$

The Venn diagram for this problem is given by the figure below

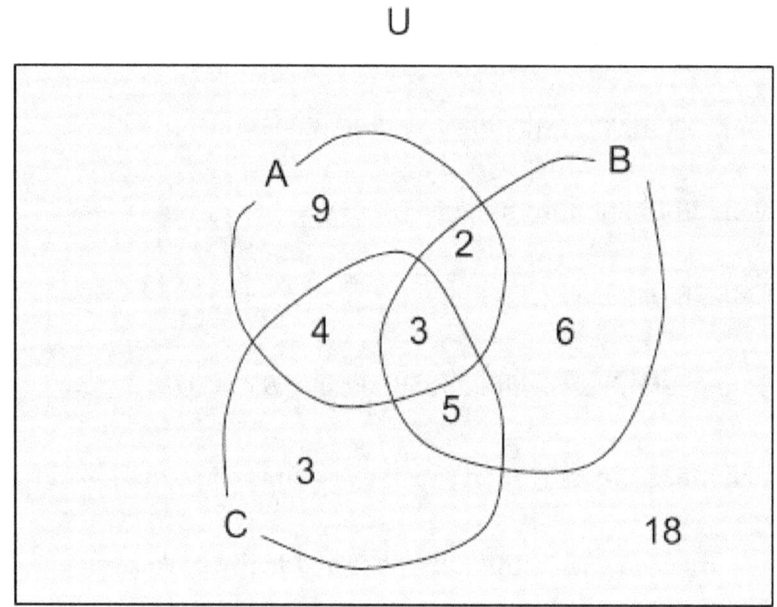

(c) **Problems requiring application of skills to familiar situations**

For instance; (i) Suppose a water tank in the shape of a right circular cylinder is thirty feet high and eight feet in diameter. How much sheet metal was used in its construction?

Solution: What you are being asked for here is to find the surface area, S, of the water tank.

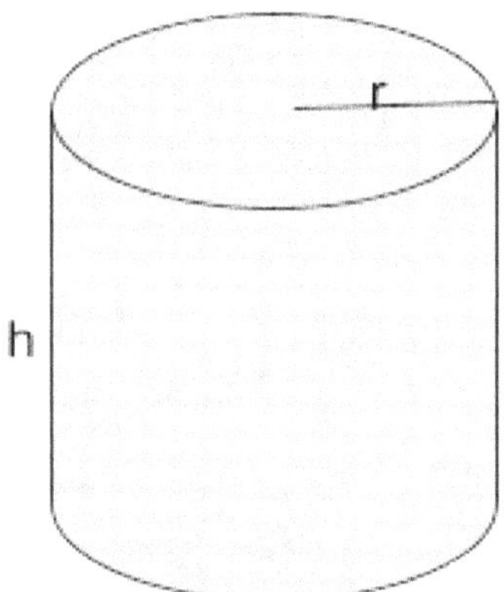

Side view of the cylindrical
tank, showing the radius, r
and height, h

The total surface area of the tank will be the sum of the surface areas of the side (the cylindrical part) and of the two ends/lids. Since the diameter is eight feet, then the radius is four feet.

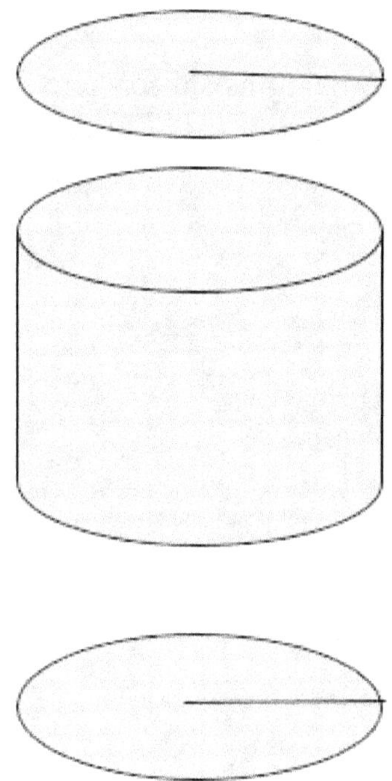

A view of the tank, showing the three separate surfaces whose areas you need to compute

The surface area of each end/lid is given by the area formula for a circle with radius r, that is, $A_1 = \pi r^2$. (There are two end/lid pieces, so you will multiply this area by 2.) The surface area of the cylinder is the circumference of the circle, multiplied by the height, that is, $A_2 = 2\pi rh$.

Therefore, the total surface area, S, of the tank is given by:

$$S = A_1 + A_2 = 2\pi r^2 + 2\pi rh = 2\pi(4)^2 + 2\pi(4)(30) = 32\pi + 240\pi = 272\pi$$

Hence, the required surface area is 272π square feet.

(ii) If the average lifespan of women in a certain country has been tracked and the equation/model for the data is $y = 0.2t + 73$ where $t = 0$ corresponds to 1960. Explain the meaning of the slope and y-intercept.

Solution: Given an equation $y = mx + c$ the slope is m and y-intercept is c, that is, the value of y when $x = 0$. In this case, the slope m = 0.2. This value tells us that, for every increase of 1 in your input variable t (that is, for every increase of one year), the value of the output variable y will increase by 0.2. What is the meaning of the slope in this case? It means that, every year, the average lifespan of women increased by 0.2 years, or about 2.4 months.

When t = 0, what is the value of y-intercept? Looking at the equation, you see that $y = 73$. What is the meaning of this y-value? It means that, in 1960 (when the counting started), the average lifespan of a woman in that country was 73 years.

(d) **Problems requiring application of skills to unfamiliar situations**

For example: (i) One morning a customer called a fencing company, wanting to fence her 1,320 square-feet garden. She ordered 148 feet of fencing, but was not asked for the width and length of the garden. You have to figure out the length and width from the given information. What are the dimensions?

Solution: The perimeter P of this rectangular area with length L and width W is given by:

$$2L + 2W = 148$$

So that

$$L + W = 74$$

The area A is given by:

$$L \times W = 1320$$

Therefore, you have the following system of equations:

$$L + W = 74$$

$$L \times W = 1320$$

You can solve either one of these equations for either one of the variables and then substitute this into the other equation. Thus, solving the first equation for L, you get

$$L = 74 - W$$

Substituting this value of L into the second equation, you have

$$(74 - W) \times W = 1320$$

This leads to a quadratic equation

$$74W - W^2 = 1320$$

Hence, you have

$$0 = W^2 - 74W + 1320$$

$$0 = (W - 30)(W - 44)$$

$$W = 30 \text{ or } W = 44$$

This gives you two valid solutions.

If W = 30, then

$$L = 74 - W = 74 - 30 = 44.$$

If W = 44, then

$$L = 74 - W = 74 - 44 = 30.$$

Therefore, the garden is 44 feet by 30 feet.

(ii) You have to make a square-bottomed, un-lidded box with a height of three inches and a volume of approximately 42 cubic inches. You do this by taking a piece of cardboard, cutting three-inch squares from each corner, scoring between the corners and folding up the edges. What should be the dimensions of the cardboard, to the nearest quarter inch?

Solution: When dealing with such type of word problems, it is usually helpful to draw a picture. Since, you will be cutting equal-sized squares out of all of the corners and the box will have a square bottom, then you know that you must start with a square piece of cardboard. You do not know how big the cardboard will be yet, so you can label the sides as having length "W" as shown below.

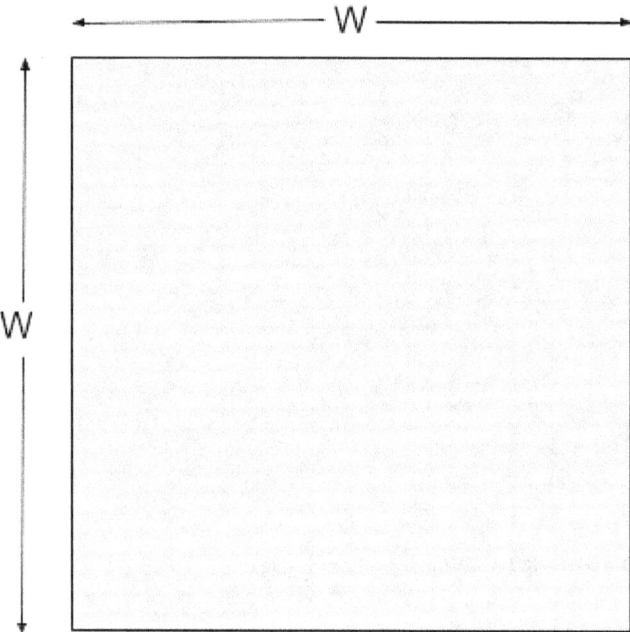

As you will be required to cut out three-by-three squares to get sides that are three inches high, you can mark that on your drawing. So, you have

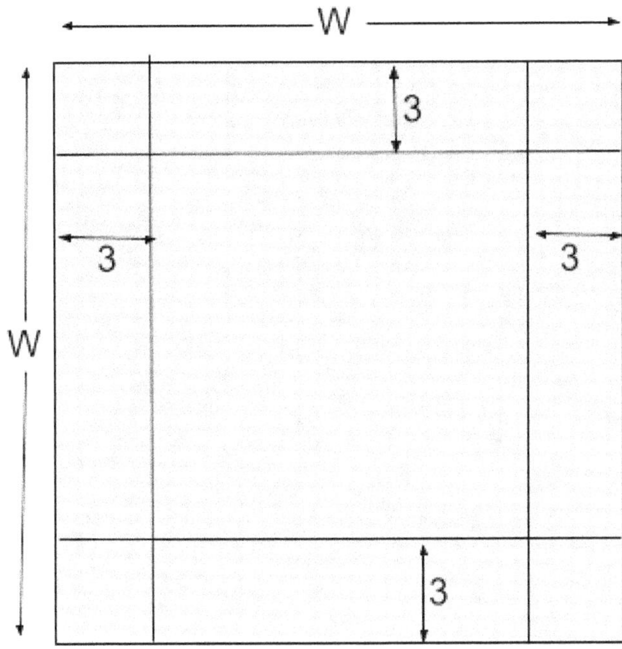

Since you will be losing three inches on either end of the cardboard when you fold up the sides, the final width of the bottom will be the original "W" inches, less three on one side and another three on the other side. That is, the width of the bottom will be $W - 3 - 3 = W - 6$.

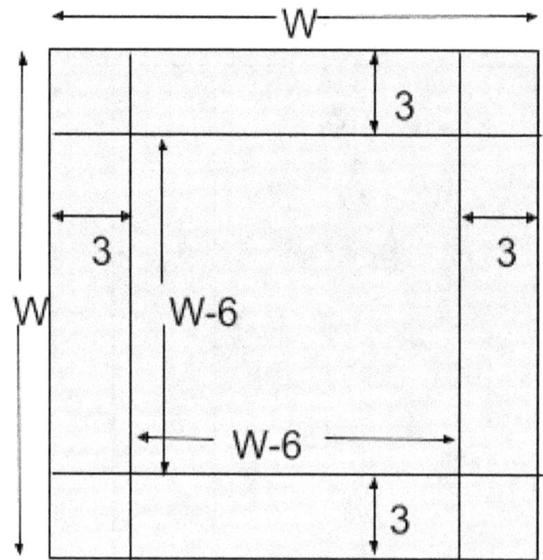

Then the volume of the box, from the drawing, is:

$$(W - 6)(W - 6)(3) = 42$$

$$(W - 6)(W - 6) = 14$$

$$(W - 6)^2 = 14$$

This is the quadratic equation you need to solve. You can take the square root of either side and then add the 6 to the right-hand side, that is,

$$W - 6 = \pm \sqrt{14}$$

$$W = 6 \pm \sqrt{14}$$

...or you can multiply out the square and apply the quadratic formula:

$$W = \frac{12 \pm \sqrt{144-88}}{2} = \frac{12 \pm \sqrt{56}}{2} = \frac{12 \pm 2\sqrt{14}}{2} = 6 \pm \sqrt{14}$$

Either way, you get two solutions which, when expressed in practical decimal terms, tell you that the width of the original cardboard is either about 2.26 inches or else about 9.74 inches.

By checking each value against the original word problem, you can know which solution value for the width is right. If the cardboard is only 2.26 inches wide, then how on earth would you be able to fold up three-inch-deep sides? But if the cardboard is 9.74 inches, then you can fold up three inches of cardboard on either side and still be left with 3.74 inches in the middle.

Checking:

$$(3.74)(3.74(3) = 41.9628$$

This is not exactly 42, but, taking the round-off error into account, it is close enough. Hence, the cardboard should measure 9.75 inches on a side.

(e)   **Problems requiring that you extend the skills or theories you know before applying them to unfamiliar situations**

For instance; (i) Suppose that there is a lady in distress who needs to be rescued. She is on a castle island surrounded by a square moat. The moat is 20 feet across and no drawbridge exists. On the shore there are two sturdy beams of wood suitable for walking across but lacking sufficient length. Each beam of wood is 19 feet long and 8 inches wide. There are no nails, screws, saws, superglue or any other method of joining the two beams to extend their length. How would you accomplish the rescue?

Solution: Thinking about variations on a situation helps you to understand which features are essential and which are unnecessary. In this case, the square shape of the moat must come into play in the solution. Looking at extremes is a potent technique of analysis in many situations and may be helpful here. The extremes, either geometrical ones as in this situation or conceptual extremes in other situations, frequently reveal features that are otherwise hidden.

Focusing attention on the corner of the moat suggests you can use one of the beams to span the corner. Of course, you need to check that the two 19-feet beams are long enough to make the configuration in the illustration below.

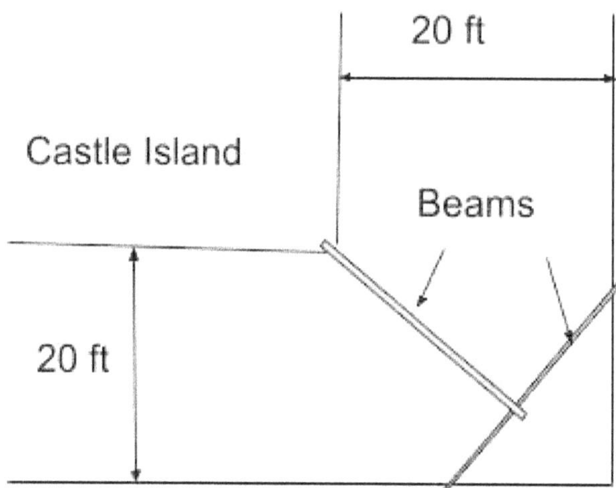

You observe that you have some right triangles. These triangles provide you with a nice opportunity to foreshadow your look at Pythagorean Theorem. You notice that the corner of the moat forms a 20-feet-by-20-feet square.

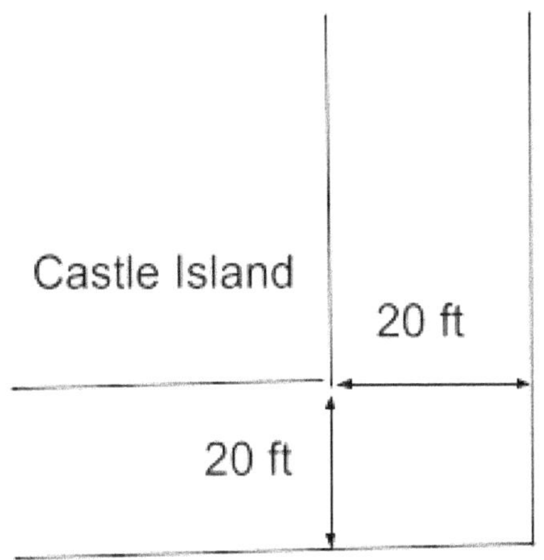

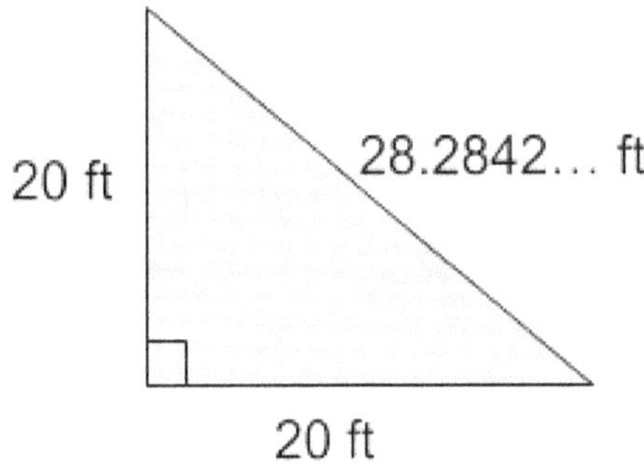

By Pythagorean Theorem, the distance from the outer corner of the shore to the other inner corner of the castle island is equal to the square root of $20^2 + 20^2$. On simplification, you see that the distance is $28.2842... feet$.

Placing the 19-feet beam diagonally across the corner of the moat as far out as it can go creates a triangle that cuts off the corner. If you draw a line from the center of the beam to the outer corner of the moat, you create two identical 45-degree right triangles, as shown below.

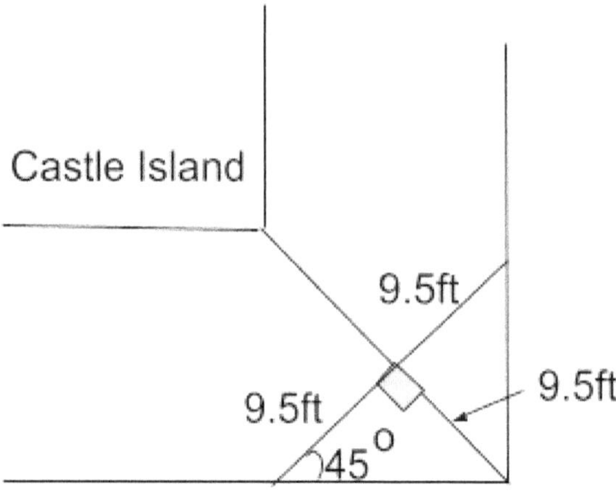

Since the length of half the beam is $9\frac{1}{2}\ feet$, you learn that the center of the beam is also $9\frac{1}{2}\ feet$ from the outer corner of the moat. As the total diagonal distance from the outer corner of the moat to the corner of the castle island is $28.2842...\ feet$, the center of the beam is $(28.2842... - 9.5)\ feet = 18.7842...\ feet$ from the corner of the castle island. Since that distance is just less than 19 feet, the other beam will just barely span the remaining distance, and the lady can be rescued.

(ii) A goat is tied to the corner of a 5-by-4-meter shed by a 8-meter piece of rope. Rounded to the nearest square meter, what is the area grazed by the goat?

Solution: What formula are you expected to use for this problem? How on earth would you answer this? When in doubt, it can be helpful to draw a picture.

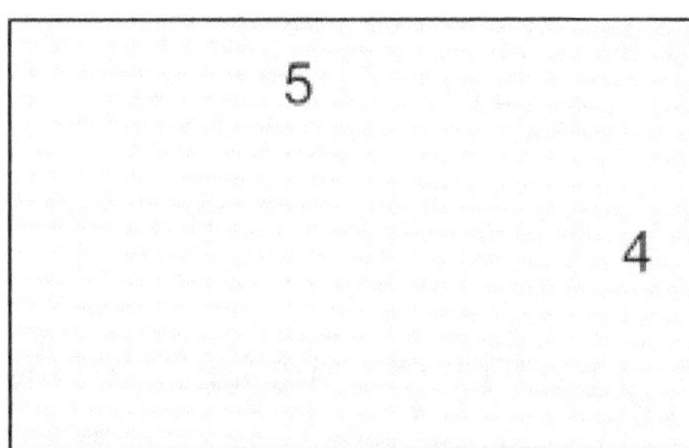

Let the goat be tied at the upper right-hand corner. Well, on the nearest two sides, the goat can go as far as is shown in the figure below:

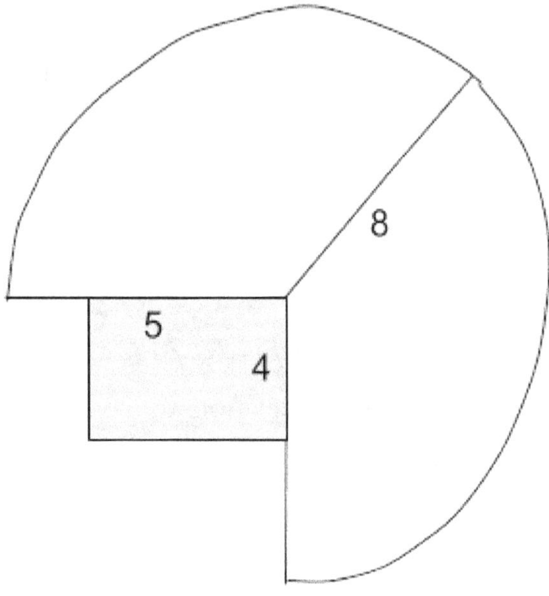

This is three-quarters of a circle with radius 8 (the full length of the rope). The goat can walk around the far corners as far as the rope will allow. Along the top side, five meters of rope will be stretched along the side, leaving another three meters "in play" while along the right-hand side, four meters will have been used, leaving another four meters:

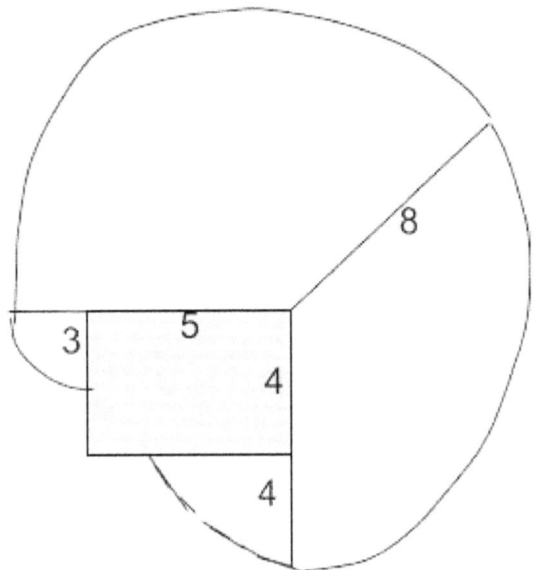

These new areas are both quarter-circles, one with radius 3 and the other with radius 4. And there will be no way for the goat to reach the opposite corner or "overlap" in grazing. So these three partial circles mark off the total grazing area.

You know the formula for the area of a circle. To find, say, the area $\frac{3}{4}$ of of a circle, you would just multiply the total area of the circle by $\frac{3}{4}$.

So, the total area grazed by the goat is:

$$\frac{3}{4}\pi(8)^2 + \frac{1}{4}\pi(3)^2 + \frac{1}{4}\pi(4)^2 = 48\pi + 2.25\pi + 4\pi = 54.25\pi$$

You are supposed to give the answer in terms of the nearest whole unit, so the goat can graze about $54.25\pi = 170 \; square\; meters$.

Note how drawing the picture allowed you to "see" what you needed to do and therefore immediately simplified your work.

4.3 Solving An Applied Problem

To solve an applied problem you:

(a)   First translate the problem into mathematics. This step is (usually) the most challenging part of an applied problem. If possible, you start by drawing a picture and labeling it with all the quantities mentioned in the problem. If a quantity in the problem is not given, you name it by a variable. You must identify the goal of the problem. Then complete the conversion of the problem into mathematics by finding equations that describe relationships among the variables.

(b) You then solve the mathematics problem you have generated using whatever suitable skills and techniques that you know.

(c)   As a final step, you should convert the answer of the mathematics problem back into words, so that you get a complete solution of the original problem.

For instance:

(i) Determine the age of Mary given that seven times her age subtracted from the square of her age gives 120.

Solution: To find the age of Mary such that when seven times her age is subtracted from the square of her age, the result is 120, you can pick a variable to stand for Mary's age. Then you can now convert the problem into a mathematics problem.

So let x be the age of Mary. Then

$$x^2 - 7x = 120$$

Looking at this equation you realize that it is a quadratic equation and so, you can employ any of the methods learnt in solving quadratic equations.

This implies that

$$x^2 - 7x - 120 = 0$$

Using factorization method, you can look for two numbers p and q such that their sum $p + q = -7$ and their product $pq = -120$. Factorizing 120, you see that its factors, both negative and positive, are $\pm 1, \pm 2, \pm 3, \pm 4, \pm 5, \pm 6, \pm 8, \pm 10, \pm 12, \pm 15, \pm 20, \pm 24, \pm 30, \pm 40, \pm 60, \pm 120$. Picking any two at a time to check which ones satisfy the given conditions, you find that only 8 and -15 do satisfy the conditions (With enough practice, you do not have to write down all the factors of 120).

Therefore, equation $x^2 - 7x - 120 = 0$ can be rewritten as

$$x^2 + 8x - 15x - 120 = 0$$

On factorizing you get

$$x(x + 8) - 15(x + 8) = 0$$

or

$$(x + 8)(x - 15) = 0$$

Thus either $x + 8 = 0$ or $x - 15 = 0$ so that $x = - 8 \; or \; x = 15$.

Similarly, you can use the quadratic equation formula

$$x = \frac{-b \pm \sqrt{b^2 - 4ac}}{2a}$$

to solve for x given any quadratic equation $ax^2 + bx + c = 0$ where $a, b \; and \; c$ are constants.

Using this quadratic equation formula to solve for x, where in your case $a = 1, b = - 7 \; and \; c = - 120$, you have

$$x = \frac{-(-7) \pm \sqrt{(-7)^2 - 4(1)(-120)}}{2(1)}$$

This equation simplifies to

$$x = \frac{7 \pm \sqrt{49 + 480}}{2} = \frac{7 \pm \sqrt{529}}{2} = \frac{7 \pm 23}{2}.$$

So that

$$x = -8 \ or \ x = 15.$$

These two methods of solving the problem led to the same answer $x = -8 \ or \ x = 15$.

Finally you have to convert the answer of the mathematics problem into words. Looking at the two values of x i.e. $-8 \ and \ 15$ we deduce that the answer should be 15 and not -8 since a person's age cannot be a negative number.

Thus the age of Mary is 15 years.

(ii) John is twice as old as his friend Peter. Peter is 5 years older than Alice. In 5 year's time, John will be three times as old as Alice. How old is Peter now?

Solution: To solve this, you employ strategies for solving a mathematical problem.

Step 1: You set up a table

	age now	age in 5 yrs
John		
Peter		
Alice		

Step 2: You fill in the table with information given in the question

Let x be Peter's age now. Using this you can get the current ages of John, Peter and Alice, respectively, as shown in the column headed "age now" in the table below. Then adding 5 to these current ages you get the respective ages in 5 year's time.

	age now	age in 5 yrs
John	2x	2x+5
Peter	x	x+5
Alice	x-5	(x-5)+5

Finally you write the new relationship in an equation using the ages in 5 yrs. You know that in 5 years, John will be three times as old as Alice. Therefore using the entries of the last column you have,

(John's age) = 3(Alice's age).

This leads you to the equation,

$$2x + 5 = 3((x - 5) + 5)$$

that simplifies to

$$2x + 5 = 3x.$$

Solving for the variable x you get

$$x = 5.$$

Therefore, the final answer is:

Peter is now 5 years old.

(iii) Twice the larger of two numbers is three more than five times the smaller. The sum of four times the larger and three times the smaller is 71. What are the two numbers?

Solution: The goal of problems like this is to give you practice in unwrapping and unwinding these words and turning the words into algebraic equations. The point is in the solving, not in the relative reality of the problem. That said, how do you solve this? The first step is to start by labeling the numbers: Let the larger number be x and the smaller one be y. You are given that twice the larger of the two numbers is three more than five times the smaller. So you have the relation 2x=5y+3. Again, the sum of four times the larger and three times the smaller is 71. Hence, you form the relation $4x + 3y = 71$. Finally, you have two equations in two variables:

$$2x = 5y + 3 \qquad (1)$$

$$4x + 3y = 71 \qquad (2)$$

You can now solve these equations simultaneously. From equation (1) you have

$$x = \frac{5}{2}y + \frac{3}{2} \qquad (3)$$

Substituting the value of x in equation (3) into equation (2) you get

$$4\left(\frac{5}{2}y + \frac{3}{2}\right) + 3y = 71$$

$$\Rightarrow 10y + 6 + 3y = 71$$

$$\Rightarrow 13y = 65$$

$$\Rightarrow y = 5$$

Now that you have the value for y, you can solve for x. This can be done by substituting the value of y into equation (3) to get

$$x = \frac{5}{2}(5) + \frac{3}{2} = \frac{25}{2} + \frac{3}{2} = 14$$

As always, you need to remember to answer the question that was actually asked. The solution here is not x = 14 and y = 5, but it is the following sentence:

The larger number is 14 while the smaller number is 5.

The trick to doing such types of problems is to label everything very explicitly. When you do that, such type of problems generally work out rather easily.

4.4 Problem Analysis

After you have worked on a problem, you must analyze it. This can help to sharpen your understanding of the problem as well as aid you when working future problems.

(a)   You should focus on the process used (not the answer) and ask yourself these questions:

(i)   What concept, formulas, and rules did you apply?

(ii)   What methods did you use?

(iii) How did you begin?

(iv) How does the solution compare with worked examples from the textbook or your notes?

(v)   Can you do this problem in another way? Can you simplify what you did?

(b)   Next, you must explain each step using your own words. You then write those explanations on a piece of paper.

(c)   Finally, you should try to think of another way of solving the same problem.

4.5 Thinking Outside The Box

You should also train yourself to be able to think outside the box, that is, be able to think critically, analytically and creatively.

For instance, try to explain a scenario in which a naked man was running after another man.

This is a very interesting situation. It poses two lines of argument. First, it could be that the naked man running after the other man was a madman. Secondly, maybe the man being chased is the one who is mad. It could have happened that the madman found the other bathing on the banks of a river and quietly put on his clothes and that is why the naked man was running after him to get back his clothes.

There is a story about a wealthy farmer who was once offered all the land he could walk on in a day, provided he came back by sunset to the point where he started. To get a head start, early the next morning the farmer started covering ground quickly because he wanted to get as much land as he could. Even though he was tired, he kept going all afternoon because he did not want to miss this once-in-a-lifetime opportunity to gain more wealth. Late in the afternoon, he realized the condition he had to fulfill to get the land was to get back to the starting point by sunset. He had gone far enough. He started his return journey, keeping an eye on how close he was to sunset. The closer it got to sunset, the faster he ran. He was exhausted, out of breath and pushed himself beyond the point of endurance. He collapsed and died before reaching the starting point. Finally, he only managed to get a six by four piece of land.

In the mathematics context, this story tells you that it is not studying a lot of mathematics that matters, what counts is how much you have understood and the level of mastery of the concepts. Mathematical rigor is like clothing: in its style it ought to suit the occasion and it diminishes comfort and restricts freedom of movement if it is either too loose or too tight. Studying mathematics is like eating food. It is not how much you eat that matters, but what counts is how much you can digest.

You should also be creative and try to have a lot of fun with mathematics. This will enhance interest and create a positive attitude towards mathematics.

Some of the funs you can have are:

(a) Take any two digit number that does not have all its digits the same. Write its digits in decreasing order, that is, from the largest to the smallest. Next, write its digits in increasing order from the smallest to the largest. Finally, subtract the smaller number from the larger one. Repeat and continue the procedure in the same way with the new number. Keep on doing this and sure enough you will end up with **9**.

Example: For the two digit number 72 you have

72	Digits in decreasing order
27	Digits in increasing order
45	Difference between them

Next, you do the same for 45.

54	Digits in decreasing order
45	Digits in increasing order
9	Difference between them

The final number is 9.

(b) Take any three digit number that does not have all its digits the same. Write its digits in decreasing order, that is, from the largest to the smallest. Next, write its digits in increasing order from the smallest to the largest. Finally, subtract the smaller number from the larger one. Repeat and continue the procedure in the same way with the new number. Keep on doing this and sure enough you will end up with **495**.

Example: For the three digit number 132 you have

321	Digits in decreasing order
123	Digits in increasing order
198	Difference between them

You do the same for 198.

981	Digits in decreasing order
189	Digits in increasing order
792	Difference between them

Next, you do the same for 792.

972	Digits in decreasing order
279	Digits in increasing order
693	Difference between them

Again, you do the same for 693.

963	Digits in decreasing order
369	Digits in increasing order
594	Difference between them

You continue in the same way for 594.

954	Digits in decreasing order
459	Digits in increasing order
495	Difference between them

Similarly, you do the same for 495.

954	Digits in decreasing order
459	Digits in increasing order
495	Difference between them

The final number is 495.

(c) Take any four digit number that does not have all its digits the same. Write its digits in decreasing order, that is, from the largest to the smallest. Next, write its digits in increasing order from the smallest to the largest. Finally, subtract the smaller number from the larger one. Repeat and continue the procedure in the same way with the new number. Keep on doing this and sure enough you will end up with **6174**.

Example: For the four digit number 1234 you have

4321	Digits in decreasing order
1234	Digits in increasing order
3087	Difference between them

You do the same for 3087.

8730	Digits in decreasing order
0378	Digits in increasing order
8352	Difference between them

Next, you do the same for 8352.

8532	Digits in decreasing order
2358	Digits in increasing order
6174	Difference between them

Again, you do the same for 6174.

7641	Digits in decreasing order
1467	Digits in increasing order

| 6174 | Difference between them |

The final number is 6174.

(d) Try to follow the following rules to multiply any two integers.

    (i) Write down two columns, headed by the two integers.

    (ii) The numbers in the left column are obtained by successively dividing by 2, ignoring remainders, until 1 is reached.

    (iii) The numbers in the right column are obtained by successively multiplying by 2, until a column of the same length, i.e. corresponding to the last number 1 in the left column, is obtained.

    (iv) Delete all even numbers in the left column, together with their corresponding numbers in the right column.

    (v) Finally, add up the numbers which remain in the right column. Their sum equals the required value of the product.

Example: To find the product 79×54, you have,

Left column	**Right column**
79	54
39	108
19	216
9	432
4	864
2	1728
1	3456

Deleting all even numbers on the left together with their corresponding numbers in the right column you get,

<u>**Left column**</u>   <u>**Right column**</u>

Left column	Right column
79	54
39	108
19	216
9	432
1	3456

Adding the numbers in the right column, you obtain,

$$54 + 108 + 216 + 432 + 3456 = 4266$$

Therefore,

$$79 \times 54 = 4266.$$

(e) Ask a friend or another student to select numbers in any two horizontal or vertical or diagonal lines on the table below, without you seeing what is being done.

1	2	3
4	5	6
7	8	9

For example, he/she could have selected 159 and 852 (it does not matter in which direction or order the numbers go). Then ask him/her to multiply these two numbers together. In this case, the product of the selected numbers is 135468. Since you have not seen what he/she has done, there is no way that you can guess what this product is. Now ask him/her to hide one of the digits, say 6 in this example. You must insist that he/she does not hold back the digit zero from the product.

Finally, ask him/her to tell you what the other digits are, in any order, apart from the hidden one. You are then able to announce immediately what the missing digit is.

How is it done? The trick relies on the fact that there are three horizontal lines, three vertical lines and two diagonals, and every three-digit number they produce is divisible by 3. Since every three-digit number which can be selected from the table has a factor 3, the product of these two numbers will contain a factor $3 \times 3 = 9$. As this product has a factor 9, the sum of the digits of the product must be divisible by 9. This is the key property to use.

When he/she reads out all the digits except the one being kept back, you add them up. You do this easily and quickly in your mind. Thus in the example, he/she tells you 13548. The sum is $1 + 3 + 5 + 4 + 8 = 21$. You know that this sum must be divisible by 9, so you can immediately say that the number he/she is holding back must be 6, since this gives $21 + 6 = 27$. Clearly 6 is the only digit which when added to 21 gives a multiple of 9. He/she should be really amazed.

Moreover, the sum of the digits does not depend upon the order in which they are read out, so he/she could scramble the digits, and tell you, for example, 14583 and you would still give the same correct answer 6.

Suppose he/she had chosen 654 and 123. Then $654 \times 123 = 80442$. If he/she holds back the digit zero, and tells you that the others are 8442, you sum these digits to obtain 18, which is divisible by 9. You then have no way of telling whether the missing digit is 0 or 9, because in both cases you will get a total sum (18 or 27) which is divisible by 9. So the trick may fail and in order to prevent this happening you tell him/her not to hide the digit zero. There can never then be any ambiguity, and the trick will always work.

(f) Request a friend or another student to write down a number consisting of three identical digits (not zero). You then tell him/her to add the three digits and divide the original number by this sum. After a few seconds, announce that he/she got 37 as the answer.

(g) Ask your friend or another student to take any number with six digits (whose digits are not all the same). Request him/her to scramble these digits, subtract the smaller number from the larger, and tell you all the digits of the difference, except one digit (not zero) which he/she hides. By adding these digits you can immediately tell him/her which digit was held back.

This stems from the fact that taking any number with several digits, scrambling these digits in any order, and subtracting the smaller from the larger, the difference is always divisible by 9 (provided that the two numbers are not identical).

(h) Request another student or your friend to divide 98765432 by 8. The result (12345679) is rather surprising, since it consists of all the digits but now in ascending order, except for 8. Next, ask him/her for his/her favorite digit – suppose it is 4. You immediately tell him/her to multiply the number 12345679 by 36 (the multiplier you supply is always the product of 9 and the stated favorite digit), and he/she will be surprised when you announce that the result consists of the digit 4 repeated nine times.

PART THREE

CHAPTER 5

STUDYING FOR A MATHEMATICS TEST

Everyday study is a big part of test preparation. Good study habits throughout the term make it easier to study for tests.

(a)   You ought to do the homework when it is assigned. You cannot hope to cram 3 or 4 weeks' worth of learning into a couple of days of study.

(b)   In tests you solve problems. Therefore, homework problems are the only way to get practice. As you do homework, you must make lists of formulas, concepts and techniques to use later when you study for tests.

(c)   You should ask your teacher questions as they arise. You must not wait until a day or two   before a test. The questions you ask right before a test should be to clear up minor details.

(d)   When studying for a test you can start by going over each section, reviewing your notes and checking that you can still do the homework problems (actually you should work the problems again). You can also use the worked examples in the textbook and notes: cover up the solutions and work the problems, and then check your work against the solutions given. You need to realize that in a textbook each problem appears at the end of the section in which you learned how to do that problem; in a test, the problems from different sections are all together.

(e)   You should ask yourself what kind of problems you have learned how to solve, what techniques of solution you have learned, how to check which techniques go with which problems and how you can use what you have leant to answer other questions that might come up.

(f)   You can try to explain out loud, in your own words, how each solution strategy is used (e.g. how to solve a quadratic equation). You can also check your verbal explanation with a friend during a study session (it is more fun than talking to yourself). This helps in that, if you get confused during a test, you can mentally return to your verbal "capsule instructions".

(g)   You must put yourself in a test-like situation: work problems from review sections at the end of chapters, and work old tests if you can find some. It is important to keep working on problems every time you are studying.

(h)   You must make sure that you start studying early: several days to a week before the test and much longer for the final test. You must begin to allot some time in your schedule for reviewing tests.

(i)   You ought to get lots of sleep the night before the test. Mathematics tests are easier when you are mentally alert/sharp.

5.1 Techniques Of Final Revision

Revision is a way of pulling your understanding together in preparation for examinations. A good revision mentality requires creativity, interactive study techniques, a high degree of motivation, time management, working well with others, good writing skills and being able to use powers of selectivity, critical thinking and memory.

The pressure of the examination stimulates you to draw together the strands of your study and to acknowledge areas that need more work. You can view this pressure negatively, as stress and the likelihood of failure, or positively as a challenge encouraging you to heighten your own expertise.

One aspect of stress is the attitude you have towards challenges. The situation and feelings that make a person panic may excite and interest another. Studying towards deadlines and examinations involve different amounts of stress for each one of you. Added life pressures, such as shortage of money, difficult relationships, bereavement or changes in your family can all add to your stress level.

Excess stress can severely affect physical and emotional health, concentration and memory. If you suffer from excess stress, you need to take steps to reduce it.

You must organize yourself to avoid stress. You can make timetables and action plans to avoid predictable crises and panic. You should take control of your time and as an examination approaches, it is useful to make adequate preparation:

(a)   You must organize your notes. The process of sorting out what is essential from what is interesting in a general way reminds you of what you have covered.

(b)   You can divide the course into quarters and begin with the last quarter, going backwards to the earlier quarters if you need to comprehend the material.

This technique has the following advantages:

(i)   Often the earlier parts of a course are there in order to prepare the way for later parts of the course, and one builds up a better understanding by beginning to study what one would like to establish.

(ii)   Often in examinations the later part of the course is over represented.

(iii) Studying in a different order makes the study more interesting.

(c)   You should collect, while studying the course, a series of very easy problems that illustrate every idea, concept and topic. This means that these problems can be done very quickly and this in itself provides a good review of the material.

(d)   You must also do old examination papers (past papers). You should do at least one of these as if it were a proper examination i.e. within the set time and without the use of aids like textbooks or notes. Past papers are your best resource. At first, the wording of examination papers can put you off and questions may seem vague as they cannot give away the answer. It is important to get used to this style well in advance of the examination. Remember that each question is a link to another area of the course. You need to find that link and consider which concepts the question is directing you to. You should look for patterns of recurring questions and find out the minimum number of topics you can revise to answer the paper.

(e)   You should also revise and learn the summary of the definitions and formulae.

Examinations are a culmination of your term's or year's learning--not just of the course content but also of strategies that you have developed over the year. Many of the strategies that help you to do well in examinations are similar to those needed for any assignment: organization, selection, developing your point of view and line of reasoning. This means that revision and examination preparation are not separate events completely divorced from the other learning activities you undertake in the year. You should remember that if you work steadily all the year, then the examination period will be more manageable.

5.2 Getting Assistance

The new material builds on the previous sections, so anything you do not understand now will make future material difficult to understand. Therefore, get help as soon as you need it. Do not wait until a test is near.

(a)   Ask questions in class. This helps you to get help and stay actively involved in the class.

(b)   Visit the teacher's office hours. Teachers like to see students who want to help themselves.

(c)   Ask friends, members of your study group, or anyone else who can help. The classmate who explains something to you learns just as much as you do, for he/she must think carefully about how to explain the particular concept or solution in a clear way. So do not be reluctant to ask a classmate.

(d)   Go to the mathematics club/help sessions or other revision sessions in school.

(e)   Find a private tutor if you cannot get enough help from other sources.

(f)   All students need help at some point, so be sure to get the help you need.

You should not be afraid to ask questions. Any question asked is better than no question at all. But a good question will allow your helper to quickly identify exactly what you do not understand.

The following will give you an insight of the kind of questions to ask.

(i)   Not too helpful comment: "I do not understand this section." This is not the best way to seek help. It does not identify what you are having trouble with and so will probably not get your question answered properly. The best you can expect in a reply to such a remark is a brief review of the section, and this will likely overlook the particular thing(s) that you do not understand. You must be specific with your questions. What exactly is it about the section that you do not understand?

(ii)   Good comment: "I do not understand why the equation of a straight line is " This is a very specific remark that will get a very specific response and hopefully clear up your difficulty.

(iii) Good question: "How can you tell the difference between the equation of a circle and the

equation of a line?"

(iv) Okay question: "How can you do question number 6?"

(v) Better question: "Can you show me how to start answering problem number 6?" (The teacher can give you a hint on how to start and let you try to finish the problem on your own), or "This is how I tried to do problem number 6. What went wrong?"

Right after you get help with a problem, you should try to work on other similar problems by yourself. Helpers are coaches, not crutches. They encourage you, give you hints as you need them and sometimes show you how to solve problems. But they should not be expected to do the work. They are there to help you figure out how to learn mathematics for yourself.

(a) When you go to the teacher's office or your study group you must have a specific list of questions prepared in advance. You should run the session as much as possible.

(b) You must not allow yourself to be sorely dependent on a tutor. The tutor cannot take the examinations for you. You must take care to be the one in control of tutoring sessions.

You must recognize that sometimes you do need some coaching to help you through, and it is up to you to seek out that coaching.

CHAPTER 6

TAKING A MATHEMATICS TEST

Just as it is important to think about how to spend your study time (in addition to actually studying), it is important to think about what strategies you must use when you take a test (in addition to actually doing the problems in the test). Good test-taking strategy can make a big difference to your grades.

The prospect of examinations can be extremely stressful, whether you have performed well or badly in the past. You may even feel resentful-that it is a waste of your time or that you know the material but cannot show your knowledge under examination conditions.

Understanding the reasons for examinations, being aware of ways that examinations can be an advantage to you and knowing that you have some control over the process can help to create the positive mindset needed for a successful examination experience.

The main purpose of an examination is for the teacher to check whether you understood what was covered in the course and that the work which demonstrates this is entirely your own.

Preparing for examinations involves a high release of energy and unusual degree of focus, which produces a very intense kind of learning. That focus and intensity are not easy to reproduce under any other conditions.

The first step to taking an examination successfully is to relax. Unfortunately, it is also one of the hardest things to do. The more worked up and nervous you are during an examination, the more likely you are to forget something or blank out. Have you ever been unable to recall information in a test situation but an hour later you were able to remember the information with clarity? This may suggest that you were not relaxed during the test. Relaxation allows more blood to flow to the brain and thus allows you to think more clearly. Your brains are alert when you are relaxed and this will enable you to perform more effectively and efficiently.

When taking a mathematics test;

(a)   You must follow the instructions that are given carefully and then face the examination with confidence. These include both the instructions written on the examination paper and those given for each question.

(b)   You should look over the entire test. This will help you to get a sense of its length. You also try to identify those problems you definitely know how to do right away and those you expect to have to think about.

(c)   You can do the problems in the order that suits you starting with the problems that you know for sure you can do. This builds confidence and means you do not miss any sure points/marks just because you run out of time. Next you try the problems you think you can figure out, and then, finally try the ones you are least sure of.

(d)   Time is of the essence. You should work as quickly and continuously as you can while still writing legibly and showing all your work. If you get stuck in a problem, you can leave it and move on to another one: you can come back to the one you were stuck with later.

"Do not let what you cannot do interfere with what you can do"

You must not waste time answering more questions than required. You ought to restrict yourself to the maximum number of questions required and do them well.

(e)   You must work by the clock. On a 50-minute, 100 point test, you have about 5 minutes for a 10 point question. Starting with the easy questions will probably put you ahead of the clock. When you work on a harder problem, you can spend the allotted time (e.g., 5 minutes) on that question, and if you have not almost finished it, you go on to another problem. You must not spend 20 minutes on a problem that will yield few or no points/marks when there are other problems still to try. You should aim at maximizing your marks.

(f)   You must make sure that you show all your work clearly, making it as easy as possible for the examiner to see how much you do know. You must try to write a well-reasoned solution. Your calculations should be arranged in a logical and clear manner showing how you proceed from one step to another. If your answer is incorrect, the examiner may assign a partial mark based on the work you show.

(g)   You should never waste time erasing. You can just draw a line through the work you want ignored and move on. Not only does erasing waste precious time, but you may discover later that you erased something useful (and/or maybe worth a partial mark if you cannot complete the problem).

(h)   In a multiple-step problem you should outline the steps before actually working the problem.

(i)   You must interpret questions given in words into mathematical statements by forming appropriate algebraic expressions. If a statement is too long, you can break it into shorter meaningful statements and then interpret each of them before solving.

(j)   You should not give up on a several-part problem just because you cannot do the first part. You can attempt the other part(s): if the actual solution depends on the first part, at least you can explain how you would do it.

(k)   You must read the questions carefully and do all parts of each problem. You should avoid making common errors due to observation. These occur due to either reading given figures or formulae wrongly or even ignoring some signs especially negative signs. Also you must avoid errors due to computation. If a question sounds like one you have done before, you have to check the wording very carefully before you select it. A slight difference in wording might require a very different answer.

(l)   While solving a problem you should make sure that you use the right mathematical symbols. For example a set $A = \{1, 2, a, g, 6\}$ but not $A = (1, 2, a, g, 6)$ or $A = (12ag6)$.

(m) It is important to verify your answers after solving the problems. You must find out whether each answer makes sense given the context of the problem.

(n)   If you finish early, you can check every problem (that means rework everything from scratch).

6.1 Test Analysis

Analyzing returned tests can aid your studying for future tests. You can ask yourself the following questions:

(a)   Did most of the test come from the lesson, textbook, or homework?

(b)   How were the problems different from those in your notes, text, and homework?

(c)   Where was your greatest source of error (careless errors, lack of time, lack of understanding of material, uncertainty of which method to choose, lack of prerequisite information, test anxiety, etc.)? You should also understand what the error was and what you did wrong. You must look for something about the error that you can remember to help you avoid making it again.

(d)   How can you change your study habits to adjust for the errors you made?

6.2 Student's Most Common Faults

Most students work hard, study examples and problems and get to the stage where they can follow most of the arguments and understand their notes and the text of their book. They can even understand the solution of typical problems. At this stage they tend to relax. But they have not learnt or understood the material well enough. They are near to being able to answer the examination questions but not quite there. Consequently, the examination proves too difficult for them, they make simple errors because they have not quite understood something or they have remembered something wrongly.

The point being made here is that you should not take anything for granted. "How would you prepare for a 100 meters race? Would you be content to get to the stage where you could run 100 meters and leave it at that?" No, of course not: you have to continue practicing regularly till the time of the race.

John, a woodcutter, worked for a company for five years but never got a pay increment. The company hired Peter and within a year he got a pay rise. This caused resentment in John and he went to his boss to talk about it. The boss said, "you are still cutting the same number of trees you were cutting five years ago. We are a result-oriented company and would be happy to give you a pay rise if your productivity goes up." John went back, started hitting hard and putting in longer hours but he still was not able to cut more trees. He went back to his boss and told him his dilemma. The boss told him to go and talk to Peter. "Maybe there is something Peter knows that you and I do not." John asked Peter how he managed to cut more trees. Peter answered, "After every tree I cut, I take a break for two minutes and sharpen my axe. When was the last time you sharpened your axe?" This question hit home like a bullet and John got his answer: that he had to regularly sharpen his axe.

Similarly, you have to sharpen your mind. You must practice, revise and practice your mathematics again and again. This will make you improve on your mastery of ideas and concepts.

Many students, while taking an examination, spoil or deface the examination paper by writing all over it. It is advisable not to do so since the question papers may be required later during revision for other examinations. It is better to scribble or write the answer triggers on the rough work page of the answer sheet. You should also make sure that you file all your past papers for future reference. You should have self-discipline and self-esteem.

A young executive with poor self-esteem was promoted but could not reconcile himself to his new office and position. One day, while he was in the office, there was a knock at his door. To show how important and busy he was, he picked up the phone and then asked the visitor to come in. As the man waited, the executive kept talking on the phone, nodding and saying, "No problem, I can handle that." After a few minutes he hung up and asked the visitor what he could do for him. The man replied, "Sir, I am here to connect your phone."

The moral of this story is that, you must not put on a study mask. Some students do pretend to be studying while their minds are far away.

Moreover, you must remember that luck favors those who try to help themselves, as the following story illustrates.

A flood was threatening a small town and everyone was leaving for safety except one man who said, "God will save me. I have faith." As the water level rose a jeep came to rescue him, the man refused, saying, "God will save me. I have faith." As the water level rose further, he went up to the second storey, and a boat came to help him. Again he refused to go, saying, "God will save me. I have faith." The water kept rising and the man climbed onto the roof. A helicopter came to rescue him, but he said, "God will save me. I have faith." Well, finally he drowned. When he reached his Maker he angrily questioned, "I had complete faith in you. Why did you ignore my prayers and let me down?" the Lord replied, "Who do you think sent you the jeep, the boat and the helicopter?"

From this story you learn that it takes action, preparation and planning rather than waiting, wondering or wishing to be successful.

Finally, you should seek God's favor and guidance. Speak to God about your wish and desire and it will be granted as long as whatever you wish to achieve is meant for His glory. Remember that you were created for a specific purpose and all your achievements must be for the fulfillment of that purpose.

During World War II, a US marine was separated from his unit on a Pacific island. The fighting had been intense, and in the smoke and the crossfire he had lost touch with his comrades. Alone in the jungle, he could hear enemy soldiers coming in his direction. Scrambling for cover, he found his way up a high ridge to several small caves in the rock. Quickly he crawled inside one of the caves. Although safe for the moment, he realized that once the enemy soldiers looking for him swept up the ridge, they would quickly search all the caves and he would be killed. As he waited, he prayed, "Lord, if it be your will, please protect me. Whatever your will though, I love you and trust you. Amen". After praying, he lay quietly listening to the enemy begin to draw close. He thought, well, I guess the Lord is not going to help me out of this one. Then he saw a spider begin to build a web over the front of his cave. As he watched, listening to the enemy searching for him all the while, the spider layered strand after strand of web across the opening of the cave. Hah, he thought. What I need is a brick wall and what the Lord has sent me is a spider web. God does have a sense of humor. As the enemy drew closer he watched from the darkness of his hideout and could see them searching one cave after another. As they came to his, he got ready to make his last stand. To his amazement, however, after glancing in the direction of his cave, they moved on. Suddenly, he realized that with the spider web over the entrance, his cave looked as if no one had entered for quite a while. "Lord, forgive me", prayed the young man. "I had forgotten that in you, a spider's web is stronger than a brick wall".

You face times of great trouble. When you do, it is so easy to forget the victories that God would work in your life, sometimes in the most surprising ways. As the great leader, Nehemiah reminded the people of Israel when they faced the task of rebuilding Jerusalem, "In God we will have success" [Nehemiah 2:20]. Remember, whatever is happening in your life, with God, a mere spider's web can become a brick wall of protection. You should believe He is with you always. Just speak His name through Jesus His son, and you will see His great power and love for you. Remember; if Christ is the center, then the radius does not matter.

PART FOUR

CHAPTER 7

SUCCESS

Everyone wants to succeed, but few people take time to study success. Similarly, everyone dislikes failure, but few people invest the time and energy necessary to learn from their mistakes.

Success is an achievement or accomplishment. It is basically about how you can turn adverse situations in your favor.

"Success is a matter of understanding and religiously practicing specific, simple habits that always lead to success"

-- Robert J. Ringer.

So, you must focus your energy in a concentrated manner onto your goal and then start perspiring for it.

Success consists of going from failure to failure without loss of enthusiasm.

"Every great improvement has come after repeated failures. Virtually nothing comes out right the first time. Failures, repeated failures, are finger posts on the road to achievement. One fails forward toward success"

-- Charles F. Kettering.

You must remember that failures are the pillars of success. They provide you with an opportunity to realize your shortcomings so that you can constantly strive to improve yourself. Failure is part of learning; you should never give up the struggle in life. If you give up during the struggle, you will never experience the victory.

The formula for success is trying until you succeed, bearing in mind that, in every effort you make lies hope for success. Remember that true success comes to those who just won't give up. It belongs only to those who are willing to work hard for it. You should make a promise to yourself that you will continue working towards your goals no matter what happens.

Life is often described as a series of events and your success is entirely dependent on choices that you make every day. Like everything that is worthwhile in life, you have to generate success on a strong foundation. You know that a great building stands on a strong foundation and so does success. The foundation of success is attitude.

Attitude is the persistent tendency to feel in a favorable or unfavorable manner towards something. It can be either positive or negative. Positive attitudes are those which tend to see the positive side of anything while negative attitudes concentrate on the negative side. In fact, any person, object, idea, or work has a mixture of both in varying proportion.

With positive attitudes, a person looks for positive phenomenon in anything but a person with negative attitudes looks for negative phenomenon in the same thing. Therefore, the phenomenon is looked at differently by two types of persons because of the difference in their attitudes.

A positive mental attitude is indispensable to your success. Think of the letters in the alphabet, **A B C D E F G H I J K L M N O P Q R S T U V W X Y Z,** represented correspondingly as numbers: **1 2 3 4 5 6 7 8 9 10 11 12 13 14 15 16 17 18 19 20 21 22 23 24 25 26.** Then, if you take three words hardwork, knowledge and attitude and sum up as percentage the numbers corresponding to their respective letters in the alphabet, you see that:

H-A-R-D-W-O-R-K gives 8+1+18+4+23+15+18+11 = **98%,**

K-N-O-W-L-E-D-G-E has 11+14+15+23+12+5+4+7+5 = **96%** but from

A-T-T-I-T-U-D-E you get 1+20+20+9+20+21+4+5 = **100%.**

This shows that your attitude is 100% of everything you do and thus attitude is very important.

You must have a positive attitude on something in order to succeed in it. If there is anything to learn that seems threatening, you should consider the ways in which it can also be an opportunity to do something new. Similarly, you should think of difficulties as challenges remembering that, the worst thing you can do when faced with challenges is to focus on them. If all you focus on is the problem you are going through, it will reproduce itself in your life.

Once upon a time a daughter complained to her father that her life was miserable and that she didn't know how she was going to make it. She was tired of fighting and struggling all the time. It seemed just as one problem was solved, another one soon followed.

Her father, a chef, took her to the kitchen. He filled three pots with water and placed each on a high fire. Once the three pots began to boil, he placed potatoes in one pot, eggs in the second pot, and ground coffee beans in the third pot. He then let them sit and boil, without saying a word to his daughter. The daughter, moaned and impatiently waited, wondering what he was doing.

After twenty minutes he turned off the burners. He took the potatoes out of the pot and placed them in a bowl. He pulled the eggs out and placed them in a bowl. He then ladled the coffee out and placed it in a cup. Turning to her he asked. "Daughter, what do you see?"

"Potatoes, eggs, and coffee," she hastily replied. "Look closer," he said, "and touch the potatoes. "She did and noted that they were soft. He then asked her to take an egg and break it. After pulling off the shell, she observed the hard-boiled egg. Finally, he asked her to sip the coffee. Its rich aroma brought a smile to her face.

"Father, what does this mean?" she asked. He then explained that the potatoes, the eggs and coffee beans had each faced the same adversity– the boiling water. However, each one reacted differently. The potato went in strong, hard, and unrelenting, but in boiling water, it became soft and weak. The egg was fragile, with the thin outer shell protecting its liquid interior until it was put in the boiling water. Then the inside of the egg became hard.

However, the ground coffee beans were unique. After they were exposed to the boiling water, they changed the water and created something new. "Which one are you," he asked his daughter. "When adversity knocks on your door, how do you respond? Are you a potato, an egg, or a coffee bean?"

The moral of the story is that: in life, things happen around you, things happen to you, but the only thing that truly matters is what happens within you. Now, which one are you?

When faced with problems, you should focus on the possible solutions. As you focus on solutions you must remember that what you behold is what you become. You should not give any situation the power to rob you of your dream.

We all know the story of David and Goliath. Goliath used to bully and harass the Israelites. One day, a 17-year old shepherd boy, David, while visiting his brothers, asked, "Why don't you stand up and fight the giant?" The brothers were terrified and they replied, "Don't you see he is too big to beat?" But David said, "No, he is not too big to beat, he is too big to miss." The rest is history. We all know what happened. David killed the giant with a sling.

So one's attitude determines how one looks at a setback. To a positive thinker, it can be a stepping stone to success, while, to a negative thinker, it can be a stumbling block. Some students think that mathematics is too difficult to pass, but, believe it or not, mathematics is too difficult to fail.

Challenges are a constant in life. They come to the king and to the street cleaner. Though they come in different measures and proportions, no one can escape facing situations that task their endurance. Unfortunately, many people allow their problems to define them. They get so engrossed in what they are going through that they lose their vision of where they are going to.

You should look beyond the challenges remembering that challenges are things you go through, not go to. They always come to pass and not to stay. In the midst of the storm, you should look ahead and let the possibilities of the future inspire you out of the conditions of the present. You must take your bearing not from where you are but from where you are going. Sometimes where you are may not inspire you, but you can seek inspiration from where you want to be and use that inspiration to shorten your stay where you are.

No one can keep you down if you insist on getting up. People stay down because they refuse to get up. No matter where you are now, you should get up and reach for more. Success is a journey and not a destination. You are constantly growing and improving and therefore a place called success, for the progressive mind, is an illusion.

If you are to succeed in mathematics, it is very important that you believe that success is possible and decide on what success means to you. You cannot change the past, but if you decide where you want to go in the future you can give yourself the best chance of getting there. You must always aim high. You must embrace a positive attitude and influence your life towards success. You must be aware that your life today is the result of your attitudes and choices you made in the past. Your life tomorrow will be the result of your attitudes and the choices you make today.

7.1 Steps To Build And Maintain A Positive Attitude

No one is born with a positive or negative attitude. Building and maintaining a positive attitude can be learnt by anyone. It may require a little direction, but with personal effort and time, you can achieve and maintain the attitude of success and leave negativity behind.

The key for building a positive attitude is to always go on, no matter what. It is not easy at times, but being firm when life becomes tough helps you to build a resilient positive attitude. Building a positive attitude is a matter of a day by day experience. Having a positive attitude does not happen overnight, but it grows as you put effort into maintaining a positive attitude daily.

Maintaining positive attitude allows you to focus on your strengths and achievements, instead of allowing negative thinking to stop you in your tracks.

To build and maintain a positive attitude, you need to practice the following steps:

7.1.1 Change focus; look for the positive

When you go digging for a kilogram of gold, you have to move tons of dirt to get a kilogram of gold. But when you go digging, you do not go looking for the dirt, you go looking for the gold. When studying mathematics you should not look at what you have covered but concentrate on what you will achieve.

An ancient Indian sage (wise man) was teaching his disciples the art of archery. He put a wooden bird on a tree as a target and asked them to aim at the eye of the bird. The first disciple was asked to describe what he saw. He said, "I see the trees, the branches, the leaves, the sky, the bird and the eye." The sage told the disciple to wait. Then he asked the second disciple the same question and he replied, "I only see the eye of the bird." The sage said, "Very good, then shoot." The arrow went straight and hit the eye of the bird.

The moral of this story is that unless you focus, you cannot achieve your goal. Whatever challenges you face, you must focus on the future and what you want to do rather than on the past. You should get a clear mental image of your ideal successful future and then take whatever action you can to begin moving in that direction.

You carry past experiences that shape you into what you are today. Crying is good if you feel pain, but you should not let the pain to dominate your life, if you do, your life will be transformed into never-ending fear and phobia. If something embarrassing happens, or happened to you, you must find a way to mitigate the effects. You should not let it continue to dominate your life and destroy your future. If you had bad experiences in the past, it does not mean that you are going to have bad experiences in your future. Your past life does not determine your future; it is what you do today that will determine your future.

As the New Testament says, "Let the dead bury the dead" you must let the past take care of itself and get your mind, your thoughts and your mental images onto the future. Whenever you are faced with a problem/difficulty, you should focus on the solution rather than on the problem. You must think and talk about the ideal solution to the obstacle or setback, rather than wasting time going over and reflecting on the problem.

Solutions are inherently positive, whereas problems are naturally negative. The instant you begin thinking in terms of solutions, you become a positive and constructive human being. You have to assume that something good is hidden within each difficulty or challenge and that whatever situation you are facing at the moment is exactly the right situation you need to ultimately be successful. You can be as positive as you want to be if you simply take actions consistent with achieving your goals rather than actions that cause you to feel the negative emotions of worry, doubt, anger and fear.

Monkey-hunters in India use a box with an opening at the top, big enough for the monkey to slide its hand in. Inside the box they put the monkey's irresistible favorite food, the nuts, then tie the box to a tree or stone and wait. The monkey grabs the nuts and its hand becomes a fist. The monkey tries to get its hand out but the opening is big enough for the hand to slide in, but too small for the fist to come out. Now the monkey has a choice, either to let go off the nuts and be free forever or hang on to the nuts and get caught.

Guess what it picks every time? You guessed it. The monkey hangs on to the nuts and gets caught.

You hang on to some nuts that keep you from going forward in mathematics. You keep on rationalizing by saying, "I cannot do this because…" and whatever comes after "because" are the nuts that you are hanging on which are holding you back. You have to let go of the nuts in order to build a positive attitude towards mathematics.

7.1.2 Make a habit of doing it now

You should never leave till tomorrow that which you can do today. You should fight procrastination by adopting "do it now." You must decide what you are going to do with every input in your life as soon as you encounter it. You must also learn to make bold decisions even when you are not really sure. A completed task is fulfilling and energizing; an incomplete task drains energy like a leak from a water tank. You must seize every opportunity that you can, get into the habit of living in the present and doing it now instead of regretting it later. You have to focus on doing what lies clearly at hand by taking the step that appears immediately in front of you. That will automatically lead to the next step, and the next, and so on, and eventually you will find yourself at your goal. You have to keep moving forward.

7.1.3 Develop an attitude of gratitude

You must count your blessings, not your troubles. Many of your blessings are hidden treasures. When you go through difficult times you can still learn to focus on the positive and be grateful for the positives in the situation. Gratitude is extremely powerful and it is important for you to learn how to be grateful in all circumstances, even when things may not look like they are going well.

By shifting the focus off from the negative aspect of a situation, you are able to see more clearly the positive things that can come out of the situation which in turn puts you into more of a vibration state that will attract the things that you desire. The bottom line is that gratitude plays an extremely important role in beginning to consciously and intentionally create desired outcomes in your life.

One day a farmer's donkey fell down into a well. The animal cried pathetically for hours as the farmer tried to figure out what to do. Finally, he decided the animal was old, and the well needed to be covered up anyway so it was not worth it to retrieve the donkey. He invited his neighbors to come over and help him. They each grabbed a shovel and began to shovel dirt into the well. At first, the donkey realized what was happening and cried horribly. Then, to everyone's amazement the donkey became quiet. A few shovel loads later, the farmer finally looked down the well. He was astonished at what he saw. With each shovel of dirt that hit his back, the donkey was doing something amazing. He would shake it off and take a step up. As the farmer's neighbors continued to shovel dirt on top of the animal, he would shake it off and take a step up. Pretty soon, everyone was amazed as the donkey stepped up over the edge of the well and happily trotted off.

This story tells you that the trick to getting out of things which bog you down is to shake them off and take a step up. Each of your troubles is a stepping stone and a blessing in disguise. You should not stop, never give up, shake it off and take a step up. After much shaking off of problems and stepping up, that is, learning from them, you will graze in green pastures.

7.1.4 Build a positive self-esteem

Self-esteem is the way you feel about yourself. When you feel well within, your performance goes up. Self-esteem is the way you talk to yourself about yourself. It has two interrelated aspects; it entails a sense of personal efficacy and a sense of personal worth. It is the integrated sum of self-confidence and self-respect. It is the conviction that you are competent to live and worthy of living.

Your self-esteem and self-image are developed by how you talk to yourself. You have conscious and unconscious memories of all the times you felt bad or wrong – they are part of the unavoidable scars of childhood. This is where the critical voice gets started. Everyone has a critical inner voice. People with low self-esteem simply have a more vicious and demeaning inner voice. You are capable of increasing your self-esteem, no matter how high or low you may feel on any given day. The good feelings you can generate about yourself are limitless. Most important, these feelings are within your control independent of how the rest of the world views you.

The first step in building self-esteem is to accept yourself now as you are before you attempt to improve. Self-acceptance and self-knowledge is the key to creating positive self-esteem. When you know and accept yourself you feel better. When you feel better, you do more. When you do more, you accomplish more. When you accomplish more, your self-confidence zooms. As your self-confidence increases, so does your self-respect. With more self-respect, your overall level of pride increases. And when you have more pride, self-respect and self-confidence, you can confidently say that you have more self-esteem. you feel more comfortable with yourself. A good, strong self-esteem shows off the best of you; a low self-esteem underscores the worst.

Nurturing your relationship with yourself is the first essential step in nurturing your relations with others. You cannot like others if you first do not like yourself, and others cannot like you, either. Self-esteem informs every relationship you have. It is not only the "real" you who live in your relationships – it is also the self-esteem within you.

The power of positive self-esteem will provide you the agility and stamina to transmute your obstacles into stepping stones as you fearlessly move through, experience and conquer what blocks so many.

Self-esteem is the foundation on which all your individually chosen thoughts, beliefs, emotions and actions are built upon with regard to yourself. It is also the foundation on how you respond and react to others and what you perceive as you encounter various events, conditions and circumstances in life.

Are you always doubtful about your capabilities? Do you often hesitate to assert yourself believing that you do not deserve a better deal? Are you incapable of setting higher goals for yourself? All of these might well be the symptoms of low self-esteem. If such is the case, then this is high time that you take a long and hard look at yourself in the mirror and take concrete steps to develop positive self-esteem.

Low self-esteem creates lots of stress and tension in your life. Because you do not value yourself highly enough you are hesitant and unsure of yourself in taking even minor decisions. You always look at others for approval and guidance. Mentally, you are prepared to be treated as a door mat, because in your heart of hearts you believe that this is what you deserve. you try to please others for getting their approval. You are always criticizing your each and every action and believe too easily in negative judgments passed on you.

Developing positive self-esteem is your prime responsibility, as in its absence you are living a half-life. Since you never know your own mind, all your decisions are influenced by those of others. Your actions do not arise out of convictions but out of a false sense of obligation to those around you. Slowly, you find yourself living a life which is false, lacking truth and thus devoid of all joy and happiness.

"To establish true self-esteem we must concentrate on our successes and forget about the failures and the negatives in our lives"

-- Denis Waitley.

To enhance the feeling of self-worth in your own eyes and awaken positive self-esteem, you must take good care of your body. You should eat food that is good for you. Do simple exercises at a convenient time or else, start going on walks. Taking care of your body gives you a good feeling, improves your image in your mind and increases your self-worth. You must treat yourself as a valuable person, dress in good clothes, maintain cleanliness and wear clothes that suit you. You must make conscious effort to look good, without being over concerned about your physical appearance.

Unfinished tasks add to the mess, the stress in your life and lower your self-image in your own eyes. Therefore, you should finish all such tasks that you have been postponing.

To nurture positive self-esteem, you must choose a group/company that appreciates you as an individual. You should also take away the focus from self by trying to get involved in the lives of those around you. You must try to share their concerns and problems by becoming a good listener.

People with low self-esteem have a habit of expecting the worst when beset with a problem. You should develop the habit of positive visualization. You have to imagine yourself successfully surmounting the problem. You must try to take a balanced view of the situation and try to control and change your initial negative reaction.

As an individual you are imperfect and thus you should accept yourself along with your deficiencies. You have to stop criticizing yourself and shift the focus from what you are not and what you do not have to what you are and what you have. It does not mean that you stop improving yourself; it means that you are acceptable as you are, but you will be even better with more efforts for self-improvement. You need to learn the art of praising yourself whenever you do something good and give yourself some incentives on successfully doing something you are proud of.

In life you are always growing and evolving. Low self-esteem takes away most of the pleasure of living and makes life a burden, full of stress and anxieties. On the other hand, positive self-esteem makes life worth living and future worth looking forward to.

7.1.5 Stay away from negative influences

Negative influences and sarcasms are all around you these days. They can take you down and wash your dreams down the drain. Negative influences not only can destroy your self-image but also pull your self-esteem down into a danger zone - never to be back again.

Some of the negative influences that you ought to stay away from are:

(i) **Negative people**

Those with whom you assemble, you soon resemble. This simple old saying hides a deep truth that can enlighten and empower every aspect of your life. Who you are - your very essence - is continually being transformed by the people you spend time with every day - family, friends, etc and the thoughts and feelings moving within you in any given moment.

An eagle's egg was placed in the nest of a hen. The egg hatched and the little eagle grew up thinking it was a chicken. The eagle did what the chicks did. It scratched in the dirt for food. It clucked and cackled. It never flew more than a few feet because that was what the chicks did. One day he saw an eagle flying gracefully and majestically in the open sky. He asked the chicks: "What is that beautiful bird?" The chicks replied, "That is an eagle. He is an outstanding bird, but you cannot fly like him because you are just a chicken." So the young eagle never gave it a second thought, believing that to be the truth. He lived the life of and died a chicken, depriving himself of his heritage because of his lack of vision. He was born to win, but conditioned to lose.

The same thing is true for you. You do not achieve excellence because of your own lack of vision. If you want to soar like an eagle, you have to learn the ways of an eagle. If you associate with achievers, you will become one. If you associate with thinkers, you will become one. If you associate with givers, you will become one and if you associate with negative people, you will become one.

"Be careful the environment you choose for it will shape you; be careful the friends you choose for you will become like them"

Negative people, generally, have the following characteristics:

- They talk in pessimistic way and emphasize that what is happening is bad, worse is yet to come.

- They criticize everything except themselves.

- They criticize the efforts and results of others, no matter how commendable they may be.

- They assign more importance to luck than efforts made to get things done.

- They derive pleasure when others are in trouble.

- They forget their blessings and count their troubles.

- They feel helpless even in small matters.

- They search for faults in everything around them.

- They try to inculcate the same attitudes which they have.

- They have a strong desire to pull others down.

Whenever you succeed in life, petty people will take cracks at you and try to pull you down. When you refuse to fight them, you win. In martial arts they teach that when someone takes a crack at you, instead of blocking, you should step away. Why? Blocking requires energy. Why not use it more productively? Similarly, in order to fight petty people, you have to come down to their level. That is what they want, because then you become one of them.

Often, you behave like others with whom you interact continuously. If you are in the company of negative people, you develop negative attitudes. Therefore, in order to protect yourself from the influence of negative people, it is necessary that you get away from such people. You should not let them to drag you down, since your character is not only judged by the company you keep but also by the company you avoid.

(ii) Smoking, drugs and alcohol

Drinking and smoking are glamorized today. It all starts with the first time. People who drink alcohol or take drugs, give a multitude of reasons, such as: to celebrate; to have fun; to forget problems; to calm down; to experiment; to make an impression on others; to socialize. But this does not help them in anyway. You should avoid smoking, drugs and alcohol as much as possible.

(iii) Pornography

Pornography is nothing except dehumanizing women and children. The consequences of pornography are that it:

- dehumanizes women

- oppresses children

- breaks marriages

- promotes sexual violence

- makes fun of ethical and moral ideals

- obliterates individuals, households and societies,

and thus it can have negative effects to your life.

(iv) **Negative television programs**

It is true that your behavior is affected by what you see around you. During childhood, you learn different things from family, friends and other people close to you that build your attitude, beliefs and even your fear as you grow up. In the current modern setting where technology is a part of your life, the media, particularly through television, plays an important part in influencing and changing your life and behavior.

Television can make those whom you are with look unimportant. The more you watch it, the more clearly it becomes that those paid to entertain you are more important than friends, neighbors and members of the family, since you pay more attention to the television.

It can stop your thinking. When you have talking heads and news anchors providing you with a constant stream of information, insights and observations, you can quite easily and comfortably find a source of opinion that allows you to stop making use of your own critical thinking.

Television can limit what you know and what you are interested in knowing. When you watch a 30 minute news cast, you often think that what you see is what there is to see. After all, if there was more to be seen, would they not show it to you and tell you more about it?

Television appears to portray or report reality, but it only gives you a small peek at what's really going on. It puts influence in the hands of a few. Television and other forms of media put just a handful of people in charge of "spoon feeding" you on news and opinion, and they have the power to slant it anyhow and anyway they want. There is selective nature of news coverage that is influenced by the few that control the editorial and the broadcasting process.

Television can dumb you down. There really are some fascinating and educational programs on television, but the overwhelming majority of television programming is designed to "spoon feed" entertainment, information and news (in that order). There is scarcely a program out there that requires you to think. Most of what you see is rapid fire information that gets little discussion to generate real rational thought. It is important to be selective on what to watch on a television.

(v) **Profanity (bad language)**

Using profanities show a lack of vocabulary, willpower, and self-control. You must change your negative language into positive language. Negative words produce negative thoughts. Negative thoughts will change your emotional output and create negative emotional feelings. These will then turn into negative decisions, which will lead to negative choices and finally become negative habits. It is a destructive cycle. Therefore, it is critical to change your mindset, avoid negative thoughts and stay focused on the positive.

(vi) **Rock music**

The lyrics of some hit songs are obscene containing bad messages and feelings that the singer wants to convey to the listeners. This can lead to undesirable behaviors, since the music you hear and the shows you watch can subconsciously influence you negatively.

7.1.6 Learn to like the things that need to be done

If you learn to like the task, the impossible becomes possible. Students work most effectively at tasks in which they are genuinely interested. Loss of interest is one of the principal causes of student failure. In every challenge you face, you must look for the valuable lesson. You must assume that every setback contains a lesson that is essential for you to learn. Only when you learn this lesson will you be smart enough and wise enough to go on to achieve the big goals that you set for yourself. Again, since you can think about only one thing at a time, when you are busy looking for the lesson then you cannot simultaneously think about the difficulty or the obstacle. Therefore, you will always find the lesson if you look for it.

7.1.7 Start each day with a positive thought

You should practice having positive thoughts and behavior daily. Just as your body needs good food every day, your mind need good thoughts every day. If you feed your body with junk food and your mind with bad thoughts, you will have both a sick body and mind. You need to feed your mind with the pure and the positive to stay on track.

Positive thinking is a mental attitude that admits into the mind thoughts, words and images that are conductive to growth, expansion and success. It is a mental attitude that expects good and favorable results. A positive mind anticipates happiness, joy, health and a successful outcome of every situation and action. Whatever the mind expects, it finds. The following two stories illustrates how this power works.

Helen applied for a new job but, as her self-esteem was low and she considered herself as a failure and unworthy of success, she was sure that she was not going to get the job. She had a negative attitude towards herself and believed that the other applicants were better and more qualified than her. Helen manifested this attitude, due to her negative past experiences with job interviews. Her mind was filled with negative thoughts and fears concerning the job for the whole week before the job interview. She was sure she would be rejected.

On the day of the interview she got up late and to her horror she discovered that the blouse she had planned to wear was dirty and the other one needed ironing. As it was already too late, she went out wearing a blouse full of wrinkles. During the interview she was tense, displayed a negative attitude, worried about her blouse and felt hungry because she did not have enough time to eat breakfast. All this distracted her mind and made it difficult for her to focus on the interview. Her overall behavior made a bad impression and consequently she materialized her fear and did not get the job.

Joyce, on the other hand, had applied for the same job too but approached the matter in a different way. She was sure that she was going to get the job. During the week preceding the interview she often visualized herself making a good impression and getting the job. In the evening before the interview she prepared the clothes she was going to wear and went to sleep a little earlier. On the day of the interview she woke up earlier than usual, had ample time to eat breakfast and then arrived to the interview venue before the scheduled time. She got the job because she made a good impression. She had also; of course, the proper qualifications for the job but so had Helen.

What do you learn from these two stories? Is there any magic employed here? No, it is all natural. When the attitude is positive you entertain pleasant feelings and constructive images and see in your mind's eye what you really want to happen. This brings brightness to your eyes, more energy and happiness. Your whole being broadcasts good will, happiness and success. Even your health is affected in a beneficial way. You walk tall and your voice is more powerful.

Your body language shows the way you feel inside. In order to turn the mind towards the positive, inner work and training are required. Attitude and thoughts do not change overnight. The power of thoughts is a mighty power that is always shaping your life. This shaping is usually done subconsciously but it is possible to make the process a conscious one.

You must ignore what others might say or think about you if they discover that you are changing the way you think. You should always visualize only favorable and beneficial situations and use positive words in your inner dialogues or when talking with others. You must endeavor to smile a little more as this helps to think positively.

You must disregard any feelings of laziness or a desire to quit. If you persevere, you will transform the way your mind thinks. Once a negative thought enters your mind, you have to be aware of it and endeavor to replace it with a constructive one. The negative thought will try again to enter your mind and then you have to replace it again with a positive one. It is as if there are two pictures in front of you and you choose to look at one of them and disregard the other.

Persistence will eventually teach your mind to think positively and ignore negative thoughts. In case you feel any inner resistance when replacing negative thoughts with positive ones you should not give up but keep looking only at the beneficial, good and happy thoughts in your mind. It does not matter what your circumstances are at the present moment, you should think positively, expect only favorable results and situations and circumstances will change accordingly. It may take some time for the changes to take place but eventually they will.

The key to success is to have no fear. You have to struggle in order to succeed. Nothing worthwhile in life comes without a struggle.

7.2 Some Factors That Make Students Succeed

Every achievement starts with a select grouping of actions that in turn create the correct environment for accomplishment. The real secret is to start with the right tools and avoid the pitfalls that those who fail–fail to do.

To succeed there are things you must have or obtain and of course conversely there are things that you either avoid or rid yourself of. There are exceptions, but the majority of successes are achieved by adherence to very key critical success factors, either knowingly or otherwise. It is important to keep in mind that there is no single best approach to success. Success arises from the interaction of many factors. What works in one situation may not necessarily transfer to success in other circumstances.

The following are some of the factors that make students succeed in mathematics:

7.2.1 Desire

The motivation to succeed comes from the burning desire to achieve a purpose. If you want something badly enough, then quitting is simply not an option. You either find a way or make one. You pay the price, whatever it takes.

One day a young man asked Socrates the secret to success. Socrates told the young man to meet him near the river the next morning. After they met, Socrates asked the young man to walk with him towards the river and they entered into the water. When the water got up to their neck, Socrates took the young man by surprise and ducked him into the water. The boy struggled to get out but Socrates was strong and kept him there until the boy started changing color. Socrates pulled his head out of the water and the first thing the man did was to gasp and take a deep breath of air. Socrates asked, "What did you want the most when you were inside the water?" The boy replied, "Air." Socrates said,

"That is the secret to success. When you want success as badly as you wanted the air, then you will get it." There is no other secret.

A burning desire is the starting point of all accomplishment. Just like a small fire cannot give much heat, a weak desire cannot produce great results. Desires become strong when they are supported by direction, dedication, determination, discipline and deadlines. You must act boldly in whatever you do as if it is impossible to fail. If you keep adding fuel to your desire, you will reach the point of knowing that you will never quit and ultimate success will be nothing more than a matter of time.

7.2.2 Commitment

Commitment ignites action. To commit is to pledge yourself to a certain purpose or line of conduct. It also means practicing your beliefs consistently. There are, therefore, two fundamental conditions for commitment. The first is having a sound set of beliefs. There is an old saying that goes, "Stand for something or you'll fall for anything." The second is faithful adherence to those beliefs with your behavior.

Possibly the best description of commitment is "persistence with a purpose." Commitment means a willingness to look for a better way and learn from the process. It focuses on eliminating complacency, confronting what is not working and providing incentives for improvement. The spirit of improving is rooted in challenging current expectation and ultimately taking the risk to make changes.

These changes are based more on optimism in the future than dissatisfaction in the past. A person who makes a commitment is willing to give up a lot. Commitment does not mean sticking to something when a person has no choice. It means sticking in spite of (regardless of) many choices. Learning to say "no" — to new commitments, to interruptions, to anything — is one of the most valuable skills you can develop to keep you focused on your own commitments and give you time to work on them.

7.2.3 Responsibility

A retiring president of a company after a standard farewell party gave two envelopes marked No. 1 and No. 2 to the incoming president, and said, "Whenever you run into a management crisis you cannot handle by yourself, open envelope No. 1. At the next crisis, open the second one."

A few years later, a major crisis came. The president went into the safe and pulled out the first envelope. It said, "Blame it on your predecessor." A few years later a second crisis came. The president went for the second envelope, and it said, "Prepare two envelopes for your successor."

What you learn from this is that, to be responsible you should accept and learn from your mistakes, remembering that a person, who never makes any mistake, never makes anything. You should not blame your failures on others or on your teachers. Your responsibility is the key to all development and learning. Success is tied to the efforts that you put into your work and the degree to which you are involved with your studies.

7.2.4 Hard work

Success is not something that you run into by accident. It takes a lot of preparation and character. Success may be described as the realization of an aim and for the realization of any aim hard work is essential. Hard work takes sacrifice and self-discipline. Hard work and practice make you better at whatever you do. Hard work helps you to develop your potential to the maximum and strive for excellence. It makes you better prepared to face adverse situations.

No matter what your dream is, you should not give up and let anything stand in your way in achieving it. Anything worth having is worth working hard for. But what it takes should not matter as long as you have the will and desire to make it happen. Giving up is easy, anyone can do that but only those who dig in deep and work hard achieve what they really want.

Life is full of twists and turns. You have to struggle here in this world to overcome every obstacle in the way to success. For this, hard work is necessary. If you see a young butterfly struggling to come out of its cocoon and you help it out, you will have robbed it of its ability to fly. This is because it is the stretching and the struggling—the process of coming out of the cocoon—that strengthens its wings to enable it to fly. Every success, therefore, must come with a process. It will be very hard for you to achieve success by just sitting idle without working hard.

Hard work helps an athlete to persevere in a race and win it. It helps an average student to become extraordinary and also it helps to transform destinies. Hard work is always followed by success.

You need to learn from nature. Like a submarine that is fast underwater, with little evidence of any movement on the surface, a duck keeps paddling relentlessly (tirelessly) underneath the water but appears smooth and calm on top. Nature gives birds their food but does not put it in their nest. They have to work hard for it. Nothing comes easy.

"If you find a path with no obstacles, it probably doesn't lead anywhere. The greater the obstacle, the more glory in overcoming it"

-- Moliere.

7.2.5 Character

Character is the will to do what is right even when it is hard. It is the sum total of a person's values, beliefs and personality.

You have to be of good character in order to succeed in mathematics. One of the things that make a huge difference between being successful and unsuccessful in mathematics is whether you are character-driven or emotion-driven.

Basically, this is the difference between the student who can make things happen and the one who waits for others to make things happen. Those who are emotion-driven tend to wait for someone else to come along and help them feel better when they are down. They may get excited for a period of time, motivated by their good feelings, but when times get tough, they lack the fortitude to continue.

Character-driven students find ways to build on the momentum and celebrate the highs and seek out small victories when spirits are down. This is because they have their eyes on things like purpose, mission and excellence. This speaks of vision. Do you realize that sometimes your emotions can get in the way of clearly seeing possibilities of the future?

When you rely too heavily on your emotions, you may find yourself paralyzed by fear, doubt or a nagging sense of needing everything to be perfect before you proceed.

Character-driven students realize that there will be good days and bad days in the attempt to do what is right. It is this belief in pursuing a right path that helps them to see their dreams come true.

The emotion-driven students may have a tendency to avoid doing mathematics problems while character-driven students move ahead in solving the problems. In fact, those who are successful in mathematics realize that nothing significant or worthwhile is accomplished without adversity in some form. Otherwise, everyone would do it.

7.2.6 Goal setting

If there is something you want to do to improve yourself, you must set reasonable goals and be prepared to be flexible if things get complicated. You must take time and use good planning measures to insure your success.

You should set goals that move you toward successful results. You fail because you do not know what you want or what you do not want. You must set S-M-A-R-T-E-R (Specific, Motivational, Action-oriented, Relevant to your situation, Time-bound,

Encouraging, Rewarding) goals that create a crystal clear roadmap for your success and work to achieve these goals every day.

You must be as specific about your goals as possible. Challenging goals are motivating. You should set goals that will push beyond what you usually think you can accomplish. You must also remember to set a deadline. A goal without a deadline is simply a dream. You should attach a realistic yet challenging deadline for accomplishment and post this where you can review it regularly.

Suppose there are two football teams enthusiastically ready to play the game, all charged up, and then someone took the goalposts away. What would happen to the game? There will be nothing left and the match will be over. To score a goal one needs to know where the goalposts are. This shows that goalposts give a sense of direction.

In mathematics context, this means that you have to find out what is expected of you then set your goals and work towards achieving them. You need to find out what you must know or be able to do by the end of the course. You need also to be aware of your own desired outcomes and keep yourself on track using short and long term goals.

It is easy to study if you have a method for working and are well organized. A skilled student uses strategies and with practice these strategies become nearly automatic. Whenever you have a goal that is unachieved, a difficulty that is unresolved, or a problem that is blocking you from getting where you want to go, with a pen and paper you must make a list of every single thing that you could possibly do to resolve the situation. The more you think on paper the more you will take control over your conscious mind. While writing down possible solutions, your mind will be positive all the time and as you write, you take advantage of the power of visualization and all kinds of insights and ideas will pop onto the page in front of you.

7.2.7 Positive believing

One of the most important things in trying to achieve any noteworthy goal is that of believing. When you expect something to happen, obstacles can then be seen not as problems but as natural events that must occur in order for the expected outcome to manifest.

Belief is complete and absolute faith and contains no doubts but only positive thinking. When you listen to your thoughts are they positive or negative? How are you programming your mind, for success or failure?

When you believe you can succeed, your mind overcomes obstacles. You solve problems creatively and are eager to take action if you believe you can succeed.

When negative thoughts enter your mind, you should recognize them for what they are and discard them. You should not be oblivious to negative feedback. Rather, you should use it constructively and refuse to let it dominate you. Consciously you must remind yourself of the positive and allow the negative to roll off your back.

Believing that you will succeed will also make you happier. Belief is on three levels: thoughts, words and actions. If you only believe something in your mind but do not demonstrate this belief through your thoughts and actions then you are going to make little progress. However when you believe on all levels, that is, if your thoughts, words and actions do not contradict each other and are all focused on the same goal then you are in control of a very powerful creative force. Indeed expectation through belief via the law of attraction moves the goal towards you and actions and thoughts move you towards the goal. Thus all three combined, can manifest your goals much quicker.

The process of simply focusing your thoughts works. How a student thinks has a profound effect on his/her performance. Have you ever had a particular thought running through your mind and then noticed a group of related subjects, events, people or places show up in your life?

Your mind is programmed through selective attention to get rid of all the clutter and leave you with only the relevant information. Having this ability to focus helps you continually discover new things, circumstances and opportunities to achieve your goals much quicker as you can continually pursue them through these new avenues with your words and actions. For example, walking down a street, an advertiser would subconsciously notice the blank advertising space on the side of the wall whereas the real estate agent would notice the vacant building nearby which could be renovated and sold. You see it is the same situation for both people but only their subconscious mind which they have programmed through their habitual thoughts allows them to notice that which they have programmed it to.

If you wish to program your mind, there are plenty of positive thinking techniques such as positive affirmations, visualization techniques, developing positive habits and so on that will allow you to do so. However, they would all ultimately come back to belief; that inner unwavering faith. So the key message here is that, there are no coincidences but only that which you bring about through your belief and focus. So it is important to believe and focus.

Those who think they can, can. There is every reason to believe that you will succeed because of all the effort you put into preparation. You must believe that your hard work will contribute to something positive. You have to believe in the best, think your best, study your best, have a goal for your best, never be satisfied with less than your best, try your best, and in the long run things will turn out for the best.

7.2.8 Persistence

Persistence means commitment and determination. Successful people set and have goals. Setting goals and being committed to achieving them fuels passion. It is that passion that powers the persistence to plow through the setbacks and navigate around the road blocks that you encounter en-route to your destination.

Have you ever watched young children learn how to walk? How many times do they get up, totter about, stand, fall, stand again, fall again, and then start to walk, one halting step at a time? How many times do they fall before they can stand and walk? It takes months of persistence for them to learn to stand and walk. They do not give up and no one would expect them to give up.

Persistence changes them into successful bipedal creatures that eventually walk, run, and engage in all manner of athletic activities. This is so because they did not give up and again because they were magnificently persistent.

Athletes put in years of practice for a few seconds or minutes of performance. They do well in the race because of persistence.

"It's not what we do once in a while that shapes our lives. It's what we do consistently"

--Anthony Robbins.

"Success is not a matter of luck or genius. Success is the outcome of adequate preparation and indomitable determination" -- Alcott.

You know the story of tortoise and hare. Hare used to brag about his speed and challenged the tortoise to a race. The tortoise accepted the challenge and they appointed the fox as the judge who gave them the starting and finishing points. The race started and the tortoise kept on going steadily. The hare ran quickly, left the tortoise behind and decided to take a nap since he was so confident he would win the race. By the time he woke up, remembered the race and started running, he saw that the tortoise had already reached the finish line and won. This story tells you that you have to be disciplined in order to work consistently and steadily.

Persistence is a commitment to finish what you start. You should program your subconscious mind for persistence well in advance of the setbacks and disappointments that you expect to have on your upward quest towards success. You must resolve in advance that you will never give up, no matter what happens.

"Many of life's failures are people who did not realize how close they were to success when they gave up"

Every time you feel like giving up, you must tell yourself to keep going. You must keep the picture of final achievement of your goal in your mind.

As a man was passing some elephants, he suddenly stopped, confused by the fact that those huge creatures were being held by only a small rope tied to their front leg. No chains, no cages. It was obvious that the elephants could, at anytime, break away from their bonds but for some reason, they did not. He saw a trainer nearby and asked why those animals just stood there and made no attempt to get away. "Well," trainer said, "when they are very young and much smaller we use the same size of rope to tie them and, at that age, it's enough to hold them. As they grow up, they are conditioned to believe they cannot break away. They believe the rope can still hold them, so they never try to break free." The man was amazed. Those animals could at any time break free from their bonds but because they believed they could not, they were stuck right where they were.

Like the elephants, how many of us go through life hanging onto a belief that we cannot do something, simply because we failed at it once before?

7.2.9 Pride of performance

Pride comes from within, which is what gives the winning edge. Pride of performance does not represent ego. It represents pleasure with humility.

Three people were laying bricks and a passerby asked them what they were doing. The first one replied, "Don't you see I am making a living?" The second one said, "Don't you see I am laying bricks?" The third one said, "I am building a beautiful monument."

Three people doing the same thing gave totally different replies. The question is: did they have different attitudes? And would their attitudes affect their performance? The answer is a clear yes.

One day a baby mosquito came back to their hiding place after flying for the first time. Its father asked "How do you feel?" The young mosquito replied "It was wonderful, everyone was clapping for me." The baby mosquito was cheerful of its performance. It looked at the clapping positively in spite of the danger it posed to its life.

These two stories teach you that, to have a good performance in whatever you do, you have to focus all your attention to the final set goal.

7.2.10 Willingness to be a student

As a student you must be willing to participate in routine school activities, such as attending class, submitting required work and following teachers' directions in class. You have to show respect to others and be willing to learn new ideas and techniques. You must be willing to listen and do homework. Moreover, as a student, you are supposed to be FAT, that is, Faithful, Available and Teachable.

You must be faithful to your teachers, study group members, parents/guardians and to other commitments.

There is a story of an ageing king who woke up one day to the realization that should he drop dead, there would be no male in the royal family to take his place. He was the last male in the royal family in a culture where only a male could succeed to the throne - and he was aging. He decided that if he could not give birth to a male, he would adopt a son who then could take his place but he insisted that such an adopted son must be extraordinary in every sense of the word. So he launched a competition in his kingdom, open to all boys, no matter what their background. Ten boys made it to the very top. There was little to separate those boys in terms of intelligence and physical attributes and capabilities. The king said to them, "I have one last test and whoever comes top will become my adopted son and heir to my throne." Then he said, "This kingdom depends solely on agriculture. So, the king must know how to cultivate plants." After that he gave a seed of corn to each boy and told them, "Take it home, plant and nurture it for three weeks. At the end of three weeks, we shall see who has done the best job of cultivating the seed. That person will be my heir-apparent." The boys took their seeds and hurried home. They each got a flower pot and planted the seed as soon as they got home. There was much excitement in the kingdom as the people waited with bated breath to see who was destined to be their next king. In one home, the boy and his parents were almost heartbroken when after days of intense care, the seed failed to sprout. He did not know what had gone wrong with his. He had selected the soil carefully, applied the right quantity and type of fertilizer, had been very dutiful in watering it at the right intervals, he had even prayed over it day and night and yet his seed had turned out to be unproductive. Some of his friends advised him to go and buy a seed from the market and plant it. "After all," they said, "how can anyone tell one seed of corn from another?" But his parents who had always taught him the value of integrity reminded him that if the king wanted them to plant any corn, he would have asked them to go for their own seed. "If you take anything different from what the king gave you that would be dishonesty. Maybe we are not destined for the throne. If so, let it be but don't be found to have deceived the king," they told him. The d-day came and the boys returned to the palace each of them proudly exhibiting a very fine corn seedling. It was obvious that the other nine boys had had great success with their seeds. The king began making his way down the line of eager boys and asked each of them, "Is this what came out of the seed I gave you?" And each boy responded, "Yes, your majesty." And the king would nod and move down the line. The king finally got to the last boy in the line-up. The boy was shaking with fear. He knew that the king was going to have him thrown into prison for wasting his seed. "What did you do with the seed I gave you?" the king asked. "I planted it and cared for it diligently, your majesty, but alas it failed to sprout." the boy said tearfully as the crowd booed him. But the king raised his hands and signaled for silence. Then he said, "My people, behold your next king." The people were confused. "Why that one?" many asked. "How can he be the right choice?" The king took his place on his throne with the boy by his side and said, "I gave these boys boiled seeds. This test was not for cultivating corn. It was the test of character; a test of integrity. It was the ultimate test. If a king must have one quality, it must be that he should be above dishonesty. Only this boy passed the test. A boiled seed cannot sprout."

We live in a society that has become obsessed with success and many show success at any cost. We say the end justifies the means. It is the tragedy of life. You see, failure often is an invitation to God to show that He is all-powerful and does not need help to make you great or to bless you.

You know, sometimes God looks for people who will trust Him completely no matter what so He can show the world that it is not by might or by power but by His spirit. God sometimes ordains failure. But many seek to circumvent divinely ordained failure by resorting to dubious means. I believe that you have been given life to lead according to God's plan by living faithfully.

The Bible says that the race is not for the swift and the battle is not for the strong. So how come, that in the world the swift wins the race and the strong the battle? It is because we are refusing to remain faithful to God and refusing to allow God to be God in all things and in our affairs. Remember, boiled seed does not sprout.

You should be available at the appointed time. To be available may cost you some time. It may cost you some effort. It may cost you some tears. It may cost you some friends. It may cost you some money. But whatever the cost, it will be worth it all when you succeed.

Teachable means an open disposition toward truth from wherever it comes, with a particular interest in learning from the teacher and the other study group participants. Being teachable means you are open regarding your lack of knowledge, skills and character. You should not act like you know it all. While you may have great passion for the role in which you serve or even some experience in it, those in leadership over you most likely have much more experience and wisdom than you.

Therefore, you must be willing to receive instruction and criticism with humility, as it is needed. As a teachable student you must be prepared for challenges that will help you reach your goals. You must be willing to let go of your own way of dealing with things, your own ideas, in order to develop new ideas and techniques. By learning and trying new ideas and techniques, you gain experience, insight and can make better use of opportunities in the future. It is also good to be an interested student. Teachers like interested students.

REFERENCES

1. Barnett, S., Discrete Mathematics: Numbers and beyond, Addison Wesley Longman Limited, 1998.

2. Baumslag, B., Fundamentals of Teaching Mathematics at University Level: Imperial College Press, London, 2000.

3. Cottrell, S. M., Skills for success: the personal development planning handbook, Basingstoke: Palgrave Macmillan, 2003.

4. Courant, R. and Robbins H., What is mathematics, revised by Stewart, I., Oxford University Press, 1996.

5. Edward, B. B. and Michael S., The Heart of Mathematics: An invitation to effective thinking, Key College Publishing, 2000.

6. Franagan, K., Maximum points, minimum panic: the essential guide to surviving exams, 2nd Edn. Dublin: Marino, 1997.

7. Freeman, R. and Mead, J., How to study effectively, Cambridge: National extension college, 1991.

8. Krantz, S. G., How to teach mathematics, Second Edition, American Mathematical Society, 1999.

9. Kulbir S. S., The teaching of mathematics, Second Edition, Sterling Publishers (P) Ltd. New Delhi, 1971.

10. Mason, J. and Burton, L. and Stacey, K., Thinking Mathematically, Addison Wesley, 1982.

11. Parsons, C., How to Study Effectively, Arrow Books Limited, 1976.

12. Polya, G., How to solve it, Princeton University Press, 1945.

13. Schoenfeld, A. H., Mathematical Problem Solving, Academic Press, 1985.

14. Shiv, K., You can win: winners don't do different things, they do things differently: a step-by-step tool for top achievers, Prentice Hall, 1998.

15. Sierpinska, A., Understanding in Mathematics, Falmer Press, 1994.

16. Stella C., The study skills handbook, Second Edition, Palgrave Macmillan, 2003.

17. Stewart, I., Concepts of Modern Mathematics, Penguin Books, 1975.

18. Stewart, I., The Problems of Mathematics, Oxford University Press, 1987.

19. Thompson, A., Critical reasoning: a practical introduction, London: Routledge, 1996.

20. Williams, K., Study skills, Basingstoke: Macmillan, 1989.

www.ingramcontent.com/pod-product-compliance
Lightning Source LLC
Chambersburg PA
CBHW060158120726
48004CB00007B/1601